The Oxford Book of
Nature Writing

The Oxford Book of
Nature Writing

Edited by

Richard Mabey

Oxford　New York

OXFORD UNIVERSITY PRESS

1995

Oxford University Press, Walton Street, Oxford OX2 6DP
Oxford New York
Athens Auckland Bangkok Bombay
Calcutta Cape Town Dar es Salaam Delhi
Florence Hong Kong Istanbul Karachi
Kuala Lumpur Madras Madrid Melbourne
Mexico City Nairobi Paris Singapore
Taipei Tokyo Toronto
and associated companies in
Berlin Ibadan

Oxford is a trade mark of Oxford University Press

British Library Cataloguing in Publication Data
Data available

Library of Congress Cataloging in Publication Data
The Oxford book of nature writing / edited by Richard Mabey.
    p.   cm.
Includes index.
    1. Natural history.   2. Nature.
I. Mabey, Richard, 1941–   .
QH81.097 1995 508—dc20 94–20026
ISBN 0–19–214172–4

10 9 8 7 6 5 4 3 2 1

Typeset by Graphicraft Typesetters Ltd, Hong Kong
Printed in Great Britain
on acid-free paper by
Bookcraft (Bath) Ltd
Midsomer Norton, Avon

Contents

Introduction

THE natural world may be the most ancient subject of human art. The vivid paintings of boars and bison discovered in caves in southern Europe pre-date writing by thousands of years. No one will ever know the real reason why they were painted. Perhaps they were the stone-age equivalent of snapshots or field-guide illustrations; elements in a map, magical charms to ensure good hunting. Yet the 20,000-year-old pictographs are so sympathetically lifelike that the question is academic. We recognize immediately the impulse behind them, the fascination with life forms dramatically different from ourselves, yet plainly filled with the same capacity for movement, fear, pain, and caring. It is a perennial interest which continues to permeate art, poetry, and mythology throughout the world. Only on the rarest occasions has nature been *ignored*.

Yet 'nature writing' is a rather special case in this long history of celebration. (I am using the phrase here in a sense which excludes bald biological recording at one extreme and fictional accounts at the other.) Deliberate attempts to portray the life of nature in prose face a huge philosophical barrier. Here, after all, is language, one of the most exquisite human inventions, resonant with the structures of human consciousness, being used to describe a world about whose inner states and meanings we can know virtually nothing. It forces us to rely on external clues, on empathy, and most notoriously on anthropomorphism, the assumption that nature shares human motives and feelings. Yet attempts to sidestep this by, for example, denying any inner lives to species other than our own, or attempting to contain their behaviour within apparently objective description (that is, description based on *our* definitions and categories), can also suffer from a kind of backdoor human-centredness.

What characterizes the most convincing nature writing is a willingness to admit both the kindredness and the otherness of the natural world. Its history is thus in part a history of our views about ourselves as a species, part of the quest for the essential characteristics and boundaries of being human. What does our atavistic affection for wild places reveal about our origins? Are we still creatures of the seasons? Can there be real communication between ourselves and other species?

Of course nature writing has a more formal history too, shaped by developments in the intellectual world. One important context has been the prevailing scientific view of nature, which has been through a multitude of changes (not all in the same direction) over the past two thousand years. The clear voice of Aristotle, writing more than 300 years BC, was, as he recognized himself, a lone one. 'I found no basis prepared,' he wrote, 'no models to copy. Mine is the first step, and therefore a small one.' The Dark Ages that followed produced no new observations or understanding of nature. The little writing and study that did emerge were chiefly confused and dogmatic attempts to interpret classical texts, and there was almost no challenge, at either a religious or a secular level, to the idea that humans should have 'dominion over' the rest of natural creation. Yet the myths and fabulous creatures of the medieval bestiaries have been too easily dismissed. For a thousand years they kept alive a sense of wonder—albeit a mystified and fearful wonder—of nature; and, read sympathetically, reveal something about early views of the interconnectedness of life.

But from the seventeenth century, a series of revolutions in thinking began to undermine the long-standing conviction that the natural world had been created solely for human use. The cracks in this anthropocentrism widened with every advance in knowledge, and brought new notes of respect and intimacy into natural history writing. Travellers brought back evidence which challenged old human presumptions, and created at the same time an enthralling new mythology populated by the bizarre creatures of the Antipodes and the prodigious plants of the rain forests. The Swedish scientist Carl Linnaeus invented in the mid-eighteenth century a system for uniquely naming, and therefore making sense of, this teeming parade of new organisms—and also proved himself capable of writing delightful, romantic cameos of his favourite animals and plants. Physico-theologists such as John Ray and William Derham reconciled scientific discovery with Christian belief by analysing design in nature, and demonstrating 'the Being and Attributes of God from his Works of Creation'. And a century later Darwin suggested how the natural world might have designed itself. In our own time this long intellectual movement has come to fruition in the science of ecology. Writers such as Aldo Leopold have argued that the inextricable links between all parts of the natural world mean that its importance is an ethical as well as a scientific matter.

Yet nature writing has also developed inside the changing context of literature as a whole—and not just non-fiction. During its critical period in the eighteenth and nineteenth centuries, for instance, nature writing was greatly influenced by what was happening in fiction. Nature had always been a rich source of objective metaphors. But just as the novel began moving from the world of external action to the interior analysis characteristic of Jane Austen and George Eliot, so nature writing began to admit the feelings of the observer. Gilbert White's beautiful and lucid journals are on the cusp of this development. They are spare, vivid, precise, yet energized by intense personal feelings about nature and his home spot. By the time of the Romantics—Coleridge especially—the *reactions* of the watcher were being quite openly explored as a legitimate subject for nature writing. They, at least, were knowable in a way that nature's inner life was not.

The tradition has found room for all kinds of voices. Natural history has been classically an amateurs' arena, a forum for diarists and potterers and dedicated enthusiasts; a world where priests and farmworkers, teachers and postmen have found inspiration, relief, and sometimes even a passion that was not always obvious in their professional lives. It gave the young Scots gardener Francis Masson an opportunity to visit South Africa in the 1770s, and write some of the first warning notes about the ecological impact of European colonialism. It helped Victorian women such as Margaret Plues and Anne Pratt find an area of achievement that was still not wholly dominated by men. And in North America, from Henry Thoreau to Annie Dillard and Edward Hoagland, it provided a landscape with enough residual wildness to liberate their words from the often over-tidy categories and conventions of European nature writing. In the United States, the best nature writers are regarded simply as writers.

Yet despite the differences in national tradition and historical fashion, what strikes one most forcibly is the common ground in nature writing. From the earliest times, the same themes have been persistently returned to: nature's profusion and profligacy, the mysteries of migration, the magnanimity of adaptation, the cycle of the seasons. There are 1,000 years separating the anonymous Irish celebration of Maytime (p. 7) and George Orwell's heartening salute to springtime toads (p. 219), but their sentiments

are identical. And both embrace the single great theme that under-
lies all nature writing—our relief and pleasure at not being alone
on the earth.

Note on the Text

The pieces included in this text are all factual prose. There are
no extracts from poetry or fiction. It is also an anthology of Western writing. Oriental, Muslim, and Buddhist writing represent
entirely different traditions which deserve their own selections.

I have organized the pieces into seven sections, each one of
which corresponds to a broad historical movement (for example
Romanticism, and the Victorian cult of natural history) and is
arranged internally in rough historical order. Each of the sections
has an introduction outlining its scientific and literary context.
But in common with most non-fiction writers, nature scribes have
commented obsessively on what they are doing, and their work
provides much of its own commentary.

The Oxford Book of
Nature Writing

Out of the Dark Ages

*B*ETWEEN *Aristotle—the uniquely clear-eyed and precocious Greek philosopher who lived four centuries before Christ—and the Renaissance, almost all natural history writing was a mix of fact and fable. For most commentators, as for ordinary people, the natural world was too mystifying and too fearful to be confronted without the help of explanatory myths and consoling interpretations.*

It was the lyrical poems of the medieval Celts that marked the beginnings of a new outlook. They are remarkable for their freshness and physical revelling in nature, and for a descriptive language whose wonderfully expressive metaphors suggest they saw the natural and human worlds as intimately linked.

Yet even the mythology that flourished in the Dark Ages (and even into the nineteenth century) was not devoid of imaginative insight. The myths of toadstones, of hibernating swallows, of mandrakes screaming when they are pulled up, of conversing nightingales, seem to reflect deep-seated wonderment and feelings of affinity as much as human credulity. And the hybrid creatures and chimeras of the bestiaries, especially the poignant vegetable lamb ('with the forme of an head hanging down and feeding on the grasse round about it'), may echo our perennial fascination with the mysteries of metamorphosis and the mutual dependence of plants and animals.

Once upon a time there was a fierce war waged between the Birds and the Beasts. For a long while the issue of the battle was uncertain, and the Bat, taking advantage of his ambiguous nature, kept aloof and remained neutral. At length when the Beasts seemed to prevail, the Bat joined their forces and appeared active in the fight; but a rally being made by the Birds, which proved successful, he was found at the end of the day among the ranks of the winning party. A peace being speedily concluded, the Bat's conduct was condemned alike by both parties, and being acknowledged by neither, and so excluded from the terms of the truce, he was obliged to skulk off as best he could, and has ever since lived in holes and corners, never daring to show his face except in the duskiness of twilight.

> AESOP, 'The Birds, the Beasts and the Bat', *c.* sixth century BC; from Thomas James's 1848 edition

The nautilus (or argonaut) is a poulpe or octopus, but one peculiar both in its nature and its habits. It rises up from deep water and swims on the surface; it rises with its shell down-turned in order that it may rise the more easily and swim with it empty, but after reaching the surface it shifts the position of the shell. In between its feelers it has a certain amount of web-growth, resembling the substance between the toes of web-footed birds; only that with these latter the substance is thick, while with the nautilus it is thin and like a spider's web. It uses this structure, when a breeze is blowing, for a sail, and lets down some of its feelers alongside as rudder-oars. If it be frightened, it fills its shell with water and sinks. With regard to the mode of generation and the growth of the shell knowledge from observation is not yet satisfactory; the shell, however, does not appear to be there from the beginning, but to grow in their case as in that of other shellfish; neither is it ascertained for certain whether the animal can live when stripped of its shell.

An accurate and sympathetic account of this species of jellyfish.

With regard to the sleeping and waking of animals, all creatures that are red-blooded and provided with legs give sensible proof that they go to sleep and that they waken up from sleep; for as a matter of fact, all animals that are furnished with eyelids shut them up when they go to sleep. Furthermore, it would appear that not only do men dream, but horses also, and dogs, and oxen; aye, and sheep, and goats, and all viviparous quadrupeds; and dogs show their dreaming by barking in their sleep. With regard to oviparous animals we cannot be sure that they dream, but most undoubtedly they sleep. And the same may be said of water animals, such as fishes, molluscs, crustaceans, to wit crawfish and the like. These animals sleep without doubt, although their sleep is of very short duration. The proof of their sleeping cannot be got from the condition of their eyes—for none of these creatures are furnished with eyelids—but can be obtained only from their motionless repose.

It is from the following facts that we may more reasonably infer that fishes sleep. Very often it is possible to take a fish off its guard so far as to catch hold of it or to give it a blow unawares; and all the while that you are preparing to catch or strike it, the fish is quite still but for a slight motion of the tail. And it is quite obvious that the animal is sleeping, from its movements if any disturbance be made during its repose; for it moves just as you would expect in a creature suddenly awakened. Further, owing to their being asleep, fish may be captured by torchlight. The watchmen in the tunny-fishery often take advantage of the fish being asleep to envelop them in a circle of nets; and it is quite obvious that they were thus sleeping by their lying still and allowing the glistening under-parts of their bodies to become visible, while the capture is taking place. They sleep in the night-time more than during the day; and so soundly at night that you may cast the net without making them stir. Fish, as a general rule, sleep close to the ground, or to the sand or to a stone at the bottom, or after concealing themselves under a rock or the ground. Flat fish go to sleep in the sand; and they can be distinguished by the outlines of their shapes in the sand, and are caught in this position by being speared with pronged instruments. The basse, the chrysophrys or gilthead, the mullet, and fish of the like sort are often caught in the daytime by the prong owing to their having been surprised when sleeping; for it is scarcely probable that such fish could be pronged while awake. Cartilaginous fish sleep at times so soundly that they may be caught by hand. The dolphin and the whale, and all such as are furnished with a blow-hole, sleep with the blow-hole over the surface of the water, and breathe through the blow-hole while they keep up a quiet flapping of their fins; indeed, some mariners assure us that they have actually heard the dolphin snoring.

ARISTOTLE, *Historia Animalium*, c.344 BC

Elephants breed in that part of Affricke which lyeth beyond the deserts and wildernesse of the Syrtes: also in Mauritania: they are found also among the Aethiopians and Troglodites, as hath been said: but India bringeth forth the biggest: as also the dragons, that are continually at variance with them, and evermore fighting, and those of such greatnesse, that they can easily claspe and wind around about the Elephants, and withal tye them fast with a knot. In this conflict they die, both the one and the other: the Elephant falls downe dead as conquered, and with his heavie weight crusheth and squeaseth the dragon that is wound and wreathed about him.

Wonderful is the wit and subtiltie that dumb creatures have & how they shift for themselves and annoy their enemies: which is the only difficultie that they have to arise and grow to so great an heigth and excessive bignesse. The dragon therefore espying the Elephant when he goeth to releese, assaileth him from an high tree and launceth himselfe upon him; but the Elephant knowing well enough he is not able to withstand his windings and knittings about him, seeketh to come close to some trees or hard rockes, and so for to crush and squise the dragon between him and them: the dragons ware hereof, entangle and snarle his feet and legges first with their taile: the Elephants on the other side, undoe those knots with their trunke as with a hand: but to prevent that againe, the dragons put in their heads into their snout, and so stop their breath, and withall, fret and gnaw the tenderest parts that they find there. Now in case these two mortall enemies chaunce to reencounter upon the way, they bristle and bridle one against another, and addresse themselves to fight; but the principall thing the dragons make at, is the eye: whereby it cometh to passe, that many times the elephants are found blind, pined for hunger, and worne away, and after much languishing, for very anguish and sorrow die of their venime. What reason should a man alledge of this so mortall warre betweene them, if it be not a verie sport of Nature and pleasure that shee takes, in matching these two so great enemies togither, and so even and equall in every respect? But some report this mutuall war between them after another sort: and that the occasion thereof ariseth from a naturall cause. For (say they) the elephants bloud is exceeding cold, and therefore the dragons be wonderfull desirous thereof to refresh and coole themselves therewith, during the parching and hote season of the yeere. And to this purpose they lie under the water, waiting their time to take the Elephants at a vantage when they are drinking. Where they catch fast hold first of their trunke:

and they have not so soone clasped and entangled it with their taile, but they set their venomous teeth in the Elephants eare, (the onely part of all their bodie, which they cannot reach unto with their trunke) and so bite it hard. Now these dragons are so big withall, that they be able to receive all the Elephants bloud. Thus are they sucked drie, untill they fall down dead: and the dragons again, drunken with their bloud, are squised under them, and die both together.

> PLINY, *Historia Naturalis*, AD 77; from the first English translation by Philomel Holland, 1601

There heard I naught but seething sea,
Ice-cold wave, awhile a song of swan.
There came to charm me gannets' pother
And whimbrels' trills for the laughter of men,
Kittiwake singing instead of mead.
Storms there the stacks thrashed, there answered them the tern
With icy feathers; full oft the erne wailed round
Spray-feathered. . . .

My guess is that this scene, so tenderly described, was observed by a young Anglo-Saxon ornithologist in some year before A.D. 685, at the Bass Rock in what his present heirs call East Lothian, most probably between what we would call 20 and 27 April by our calendar. Birds change their distribution, but not so much their season; just at this stormy time the winds can blow cold in the Firth of Forth, and the great whooper swans pass north to their breeding-grounds on the moorland wetlands, and the whimbrels utter their trilling titter, usually of seven beats, on their flight to the moorland drylands; and the common terns' main arrival is due; and the gannets and the kittiwakes and the white-tailed sea eagles hold their nest-sites on the ocean-facing cliffs. At least the ernes did: they have gone now, from the Bass, though a statement of the late, great Professor Alfred Newton of a century ago —a little vague as from so deep a scholar—indicates that the white-tailed eagle still has an eyrie on the Bass in our bird-historical times (that is, since 1600) a thousand years after the Seafarer saw them there.

This dating, and placing, of perhaps the first bit of true-sounding, wild-inspired field ornithological record since the Romans gave up their colony is of course no more than an (I hope) educated guess. The Bass is the only place I can think of within the Anglo-Saxon realm in the Dark Ages where all these birds could have been seen together under the circumstances which the unknown author of *The Seafarer* describes. There is, today, it is true, a small gannet colony on the chalk ledges of Bempton Cliffs in Yorkshire, with kittiwakes, where terns and whimbrels pass. But it is quite new. Nowhere else in England save on Lundy in the Bristol Channel have gannets nested since 'Domesday Book' times; and the description of the Seafarer is no description of Lundy. The gannet colony of the Bass is one of the old, ancient ones. It was flourishing in 1447, and if a conservative, home-true bird like the gannet can throng a breeding-rock (only thirty, now (and probably never more at any one time in the past), are used by the North Atlantic gannet) for half a millennium, it can for a millennium and a half. 'Hleo ⊅or' is a very good representation of the *urrah . . . urrah* of the breeding gannet. The terns fit; to this day common terns nest on the Bass's skerry-neighbours Craigleith, Lamb and Fidra. Icy storm or no, their mantles ever have the sheen of ice. The erne fitted. The kittiwakes fit—and what a glorious bit of onomatopoiea the poet perpetrated with his 'medodrince': just what these sea-mews were doubtless heard to say in Anglo-Saxon times, just as they are heard to say, and are called, kittiwake in Britain now, kishiefaik and killyweeack in Orkney in the old days, *krykkje* in Norway, *pikkukajava* by the Finns, *tâterâq* by the Greenlanders, tickle-arse by the Newfoundlanders. The whoopers fit; they are the singing swans of the north that doubtless nested widely in the Scottish loch-side bogs in the Dark Ages, and sometimes do so even now. The whimbrel fits, though scholars still dispute whether the Anglo-Saxon *huilpe* meant the whimbrel or its close congener the curlew. It probably meant both. Whimbrels nest in the northernmost Highlands to this day; curlews all over the Scottish moors. But the trill of the whimbrel is perhaps better laughter than the curlew's musical April bubble: at least I think so. *Huilpe*, it is true, becomes the Scottish whaup for curlew (1538 earliest in the *Oxford English Dictionary*) and the Dutch *wulp* for curlew (and *regenwulp* for whimbrel). But the whimbrel whilps as well as titters, and I favour it for the Seafarer's bird; and even if his bird *was* the curlew, the Bass must still have been the Seafarer's place.

> JAMES FISHER, *The Shell Bird Book*, 1966; Fisher's commentary on his own translation of a passage from *The Seafarer*, c.680, an evocative picture of Dark Ages ornithology

May-time, fair season, perfect is its aspect then; blackbirds sing a full song, if there be a scanty beam of day.

The hardy, busy cuckoo calls, welcome noble summer! It calms the bitterness of bad weather, the branching wood is a prickly hedge.

Summer brings low the little stream, the swift herd makes for the water, the long hair of the heather spreads out, the weak white cotton-grass flourishes.

. . . The smooth sea flows, season when the ocean falls asleep; flowers cover the world.

Bees, whose strength is small, carry with their feet a load reaped from the flowers; the mountain allures the cattle, the ant makes a rich meal.

The harp of the wood plays melody, its music brings perfect peace; colour has settled on every hill, haze on the lake of full water.

The corncrake clacks, a strenuous bard; the high pure waterfall sings a greeting to the warm pool; rustling of rushes has come.

Light swallows dart on high, brisk music encircles the hill, tender rich fruits bud . . .

> ANON Irish, 'May-time', ninth–tenth century; tr. Kenneth Jackson, 1971

It happened that St Kevin, deserting the company of man as he always did in the Lenten season, lived in a hut which sufficed to protect him from the sun and the rain. As was his habit he put his hand out through the window of the hut and raised it towards heaven, when a blackbird alighted and laid its eggs there in his hand as if it were a nest. Such feeling had St Kevin for this bird that he neither drew his hand back nor closed his fingers; which he held just so, without tiring, until the young ones were hatched and ready to fly. So in all Ireland St Kevin is shown with a blackbird on his outstretched hand, in memory of this wonderful event.

> GIRALDUS CAMBRENSIS, 'How St Kevin Made a Blackbird's Nest', c.1188: version by Geoffrey Grigson, in *Rainbows, Fleas and Flowers*, 1971

The Teivi has another singular particularity, being the only river in Wales, or even in England, which has beavers; in Scotland they are said to be found in one river, but are very scarce. I think it not a useless labour, to insert a few remarks respecting the nature of these animals; the manner in which they bring their materials from the woods to the water, and with what skill they connect them in the construction of their dwellings in the midst of rivers; their means of defence on the eastern and western sides against hunters; and also concerning their fish-like tails.

The beavers, in order to construct their castles in the middle of rivers, make use of the animals of their own species instead of carts, who, by a wonderful mode of carriage, convey the timber from the woods to the rivers. Some of them, obeying the dictates of nature, receive on their bellies the logs of wood cut off by their associates, which they hold tight with their feet, and thus with transverse pieces placed in their mouths, are drawn along backwards, with their cargo, by other beavers, who fasten themselves with their teeth to the raft. The moles use a similar artifice in clearing out the dirt from the cavities they form by scraping. In some deep and still corner of the river, the beavers use such skill in the construction of their habitations, that not a drop of water can penetrate, or the force of storms shake them; nor do they fear any violence but that of mankind, nor even that, unless well armed. They entwine the branches of willows with other wood, and different kinds of leaves, to the usual height of the water, and having made within-side a communication from floor to floor, they elevate a kind of stage, or scaffold, from which they may observe and watch the rising of the waters. In the course of time, their habitations bear the appearance of a grove of willow trees, rude and natural without, but artfully constructed within. This animal can remain in or under water at its pleasure, like the frog or seal, who show, by the smoothness or roughness of their skins, the flux and reflux of the sea. These three animals, therefore, live indifferently under the water, or in the air, and have short legs, broad bodies, stubbed tails, and resemble the mole in their corporal shape. It is worthy of remark, that the beaver has but four teeth, two above, and two below, which being broad and sharp, cut like a carpenter's axe, and as such he uses them. They make excavations and dry hiding places in the banks near their dwellings, and when they hear the stroke of the hunter, who with sharp poles endeavours to penetrate them, they fly as soon as possible to the defence of their castle, having first blown out the water from the entrance of the hole, and rendered it foul and muddy by scraping the earth, in order thus artfully to elude the stratagems of the well-armed hunter, who is watching them from the opposite banks of the river. When the beaver finds he cannot save himself from the pursuit of the dogs who follow him, that he may ransom his body by the sacrifice of a part, he throws away that, which by natural instinct he knows to be the object sought for, and in the sight of the hunter castrates himself, from which circumstance he has gained the name of Castor; and if by chance the dogs should chase an animal which had been previously castrated, he has the sagacity to run to an elevated spot, and there lifting up his leg, shows the hunter that the object of his pursuit is gone. Cicero speaking of them says, 'They ransom themselves by that part of the body, for which they are chiefly sought.'. . . Thus, therefore, in order to preserve his skin, which is sought after in the West, and the medicinal part of his body, which is coveted in the East, although he cannot save himself entirely, yet, by a wonderful instinct and sagacity, he endeavours to avoid the stratagems of his pursuers. The beavers have broad, short tails, thick, like the palm of a hand, which they use as a rudder in swimming; and although the rest of their body is hairy, this part, like that of seals, is without hair, and smooth; upon which account, in Germany and the arctic regions, where beavers abound, great and religious persons, in times of fasting, eat the tails of this fish-like animal, as having both the taste and colour of fish.

> Giraldus Cambrensis, *Itinerary Through Wales*, c.1188.
> Beavers were either extinct or very rare in Wales by the twelfth century, so Giraldus can be forgiven for including, in an otherwise accurate account, the ancient myth that the animals bit off their own testicles when pursued. What early observers may have witnessed were beavers licking the scent glands at the base of their tails

CROSSBILLS

At the turn of the same year, at the season of fruits, certain wonderful birds never before seen in England appeared, particularly in orchards. They were a little bigger than larks and ate the pips of apples and nothing else from the apples. So they robbed

the trees of their fruit very grievously. Moreover they had the
parts of the beak crossed [*cancellatas*, literally 'lattice-wise'] and
with them split the apples as if with pincers or a pocket-knife.
The pieces of the apples which they left were apparently tainted
with poison.

MATTHEW PARIS, *Chronica Majora*, 1251

TO A BIRCH-TREE CUT DOWN, AND SET UP IN LLANIDLOES FOR A MAYPOLE

Long are you exiled from the wooded slope, birch-tree, with your
green hair in wretched state; you who were the majestic sceptre
of the wood where you were reared, a green veil, are now turned
traitress to the grove. Your precinct was lodging for me and my
love-messenger in the short nights of May. Manifold once (ah,
odious plight!) were the carollings in your pure green crest, and
in your bright green house I heard every bird-song make its way;
under your spreading boughs grew herbs of every kind among
the hazel saplings, when your dwelling-place in the wood was
pleasing to my girl last year. But now you think no more of love,
your crest above remains dumb; and from the green meadow and
the upland, where your high rank was plain to see, you have gone
bodily and in spite of the cost to the town where trade is brisk.
Though the gift of an honourable place in thronged Llanidloes
where many meet is good, not good, my birch, do I think your
rape nor your site nor your habitation. No good place is it for
you for putting out green leaves, there where you make grimaces.
Every town has gardens with leafage green enough; and was it
not barbarous, my birch, to make you wither yonder, a bare pole
by the pillory? If you had not come, at the time of leaves, to
stand in the centre of the dry crossroads, though they say your
place is a pleasant one, my tree, the skies of the glen would have
been the better. No more will the birds sleep, no more will they
sing in their shrill note on your fair gentle crest, sister of the
dusky wood, so incessant will be the hubbub of the people around
your tent—a cruel maiming! and the green grass will not grow
beneath you, for the trampling of the townsmen's feet, any more
than it grew on the wind-swift path of Adam and the first woman
long ago. You were made, it seems, for huckstering, as you stand
there like a market-woman; and in the cheerful babble at the fair
all will point their fingers at your suffering, in your one grey shirt

and your old fur, amid the petty merchandise. No more will the bracken hide your urgent seedlings, where your sister stays; no more will there be mysteries and secrets shared, and shade, under your dear eaves; you will not conceal the April primroses, with their gaze directed upwards; you will not think now to inquire, fair poet tree, after the birds of the glen. God! Woe to us, a cramped chill is on the land, a subtle dread, since this helplessness has come on you, who bore your head and your fine crest like noble Tegwedd of old. Choose from the two, since it is foolish for you to be a townsman, captive tree: either to go home to the lovely mountain pasture, or to wither yonder in the town.

> GRUFFYD AB ADDAF AP DAFYDD, 'To a Birch-Tree Cut Down, and Set Up in LLanidloes for a Maypole', *c.*1340–70; tr. Kenneth Jackson, 1971

Bartholomaeus Anglicus was a Minorite friar whose monumental work De Proprietatibus Rerum *included the first truly popular nature writing of the Middle Ages. Its clarity and obvious affection for the natural world mark it out from almost everything written in Europe in the previous few centuries. But it has none of the vivid attention to detail of Albertus Magnus, a Dominican friar and follower of Aristotle from Swabia, whose* De Vegetabilibus *stands as the first text of Renaissance natural history.*

Woods are wide places waste and desolate where many trees grow in without fruit and also few having fruit. And those trees which are barren and bear no manner fruit always are generally more and higher than those with fruit, few out taken as oak and beech. In these woods are often wild beasts and fowls. Therein grow herbs, grass, lees and pasture, and namely medicinal herbs in woods are found. In summer woods are beautied with boughs and branches, with herbs and grass. In wood is place of deceit and of hunting. For therein wild beasts are hunted: and watches and deceits are ordained and let of hounds and of hunters. There is place of hiding and of lurking. For often in woods thieves are hid, and often in their awaits and deceits passing men come and are spoiled and robbed and often slain. And so for many and diverse ways and uncertain men often err and go out of the way. And take uncertain way and the way that is unknown before the way that is known and come often to the place these thieves lie in wait and not without peril. Therefore are often knots made on trees and in bushes in boughs and in branches of trees; in token and mark of the highway; to show the certain and sure way to wayfaring men. But often thieves in turning and meeting of ways change such knots and signs and beguile many men and bring them out of the right way by false tokens and signs. Birds, fowls and bees flee to woods, birds to make nests and bees to gather honey. Birds to keep themself from fowlers and bees to hide themselves to make honeycombs privily in hollow trees and stocks. Also woods for thickness of trees are cold with shadow. And in heat of the sun weary wayfaring and traveling men have liking to have rest and to heal themself in the shadow. Many woods are between divers countries and lands: and depart them asunder. And by weaving and casting together of trees often men keep and defend themself from enemies.

> BARTHOLOMAEUS ANGLICUS, *De Proprietatibus Rerum*, late fourteenth century; original translation by the Cornishman John de Trevisa, 1398

The oak is a very big, tall and wide-branched tree with many big and deep roots and a bark, rough when it gets older, but soft in its youth. It possesses very big branches. The leaf is thick, broad and hard, when it has reached full strength, and can be entirely circumscribed by triangles with their bases on the leaf and their apexes outside. The leaf keeps to the tree for a long time, even when it is dry. The wood grows out of upright layers with upright pores, can easily be cut perpendicularly, is easily hewable, and suitable for figure-carving. For this purpose, however, boxwood is better still. The outside layers of the timber are white, towards the inside they merge into red. In water it first floats, but later sinks, because of its terrestriality, and gets black. . . . Its fruit is called the acorn. It is not attached to the twig on which it grows by its own little bract, but little cups develop on the twig, and in these the acorns sprout forth. The acorn has outside a hard pod which holds it. It is like wood, very smooth and of a columnar shape, except that it is not flat at the top, but ends hemispherically with a point in the middle like the mark of the pole. At the bottom it has its base with which it sucks nourishment out of its cup. This base also has not a completely plain surface, but is a kind of hemisphere considerably flattened at the pole. It is in contact with the place of nourishment, but not merely in a point. Otherwise it could not draw sufficient nourishment. The acorn in its pod has a skin which is not hard but soft. This develops out of the husk of the acorn, and the acorn is wrapped up in it and has a division along the middle, as though a column had its surface cut throughout its length. On the top it has the seed, and what appears of a mealy substance below is purified into food for the seed. The cup in which the acorn itself sits is concave, well formed, almost as if it were turned. Its bottom is somewhat flat, and sitting in this the acorn draws its food. The outside of the cup is scaly and rough because of its terrestriality, which has been purified from the substance of the acorn. It is not connected with the twig either by a little bract or a hanging stalk. Instead of that it sits immediately on the twig. The reason of this is that the acorn should not be too far away from the twig, because if the acorn had to suck its nourishment up a long distance this would get hard and cold on the way and be of no use, especially since the juice of the oak is very terrestrial. On the leaves of the oak you can often find a round growth like a ball which is called the oak-gall. If this has been there for some time it brings forth a maggot.

ALBERTUS MAGNUS, *De Vegetabilibus*, c.1250–75; tr. Borgnet, 1891

A TENTH-CENTURY CHINESE CLASSIFICATION OF THE ANIMAL WORLD

(1) Those Belonging to the Emperor; (2) Embalmed; (3) Tame; (4) Suckling Pigs; (5) Sirens; (6) Fabulous; (7) Stray Dogs; (8) Included in the Present Classification; (9) Frenzied; (10) Innumerable; (11) Drawn with a Very Fine Camelhair Brush; (12) *Et Cetera*; (13) Having Just Broken the Water Pitcher; and (14) That From a Long Way Off Look Like Flies.

TAI PING KUANG CHI, AD 981; tr. Jorge Luis Borges. The French philosopher Michel Foucault writes of this classification: 'In the wonderment of this taxonomy, the thing we apprehend in one great leap, the thing that, by means of the fable, is demonstrated as the exotic charm of another system of thought, is the limitations of our own . . .' (*Les mots et les choses*)

Amongst the curious myths of the Middle Ages none were more
extravagant and persistent than that of the 'Vegetable Lamb of
Tartary,' known also as the 'Scythian Lamb,' and the 'Borametz,'
or 'Barometz,' the latter title being derived from a Tartar word
signifying 'a lamb.' This 'lamb' was described as being at the
same time both a true animal and a living plant. According to
some writers this composite 'plant-animal' was the fruit of a tree
which sprang from a seed like that of a melon, or gourd; and
when the fruit or seed-pod of this tree was fully ripe it burst open
and disclosed to view within it a little lamb, perfect in form, and
in every way resembling an ordinary lamb naturally born. This
remarkable tree was supposed to grow in the territory of 'the
Tartars of the East', formerly called 'Scythia'; and it was said
that from the fleeces of these 'tree-lambs,' which were of surpassing whiteness, the natives of the country where they were
found wove materials for their garments and 'headdress.' In the
course of time another version of the story was circulated, in
which the lamb was not described as being the fruit of a tree,
but as being a living lamb attached by its navel to a short stem
rooted in the earth. The stem, or stalk, on which the lamb was
thus suspended above the ground was sufficiently flexible to allow
the animal to bend downward, and browze on the herbage within its reach. When all the grass within the length of its tether had
been consumed the stem withered and the lamb died. This plant-
lamb was reported to have bones, blood, and delicate flesh, and
to be a favourite food of wolves, though no other carnivorous
animal would attack it.

HENRY LEE, *The Vegetable Lamb of Tartary*, 1887

HOW NEW KINDS OF PLANTS MAY BE GENERATED OF
PUTREFACTION

As we have showed before, that new kinds of living creatures may
be generated of putrefaction; so, to proceed in the same order as
we have begun, we will now show that new kinds of plants may
grow up of their own accord, without any help of seed or such
like. The Ancients questionless were of opinion, that divers plants
were generated of the earth and water mixed together; and that
particular places did yield certain particular plants . . .

I my self have oft-times by experience proved, that ground
digged out from under the lowest foundations of certain houses,

and the bottom of some pits, and laid open in some small vessel to the force of the sun, hath brought forth divers kinds of plants. And whereas I had oftentimes, partly for my own pleasure, and partly to search into the works of Nature, sought out and gathered together earths of divers kinds, I laid them abroad in the sun, and watered them often with a little sprinkling, and found thereby, that a fine light earth would bring forth herbs that had slight stalkes like a rush, and leaves full of fine little rags; and likewise that a rough and stiff earth full of holes, would bring forth a slight herb, hard as wood, and full of crevices. In like manner, if I took of the earth that had been digged out of the thick woods, or out of moist places, or out of the holes that are in hollow stones, it would bring forth herbs that had smooth bluish stalkes, and leaves full of juice and substance, such as Pennywort, Purslane, Fenugreek, and Stonecrop. We made trial also of some kinds of earth that had been far fetched, such as they had used for the ballast of their ships; and we found such herbs generated thereof, as we knew not what they were. Nay further also, even out of very roots and barks of trees, and rotten seeds, powned and buried, and there macerated with water, we have brought forth in a manner the very same herbs; as out of an oaken root, the herb Polypody, and Oak-fern, and Spleenwort, or at least such herbs as did resemble those, both in making and in properties. What should I here rehearse, how many kinds of toad-stools and puffs we have produced? yea, of every several mixture of putrified things, so many several kinds have been generated. All which I would here have set down, if I could have reduced them into any method; or else if such plants had been produced, as I intended: but those came that were never sought for. But happily I shall hereafter, if God will, write of these things, for the delight, and speculation, and profit of the more curious sort: which I have neither time nor leisure not to mention, seeing this work is rustled up in haste.

HOW PLANTS ARE CHANGED, ONE OF THEM DEGENERATING INTO THE FORM OF THE OTHER

To work miracles, is nothing else (as I suppose) but to turn one thing into another, or to effect those things which are contrary to the ordinary course of Nature. It may be done by negligence, or by cunning handling and dressing them, that plants may forsake their own natural kind, and be quite turned into another kind; wholly degenerating, both in taste, and colour, and bigness, and fashion: and this I say may easily be done, either if you neglect to dress or handle them according to their kind, or else dress them more carefully and artificially than their own kind requires. Furthermore, every plant hath his proper manner, and peculiar kind of sowing or plants: for some must be sowed by seed, others planted by the whole stem, others set by some root, others grafted by some sprig or branch: so that if that which should be sowed by seed, be planted by the root, or set by the whole stock, or grafted by some branch; or if any that should be thus planted be sowed by seed; that which cometh up will be of a divers kind from that which grows usually, if it be planted according to its own nature, as Theophrastus writes. Likewise if you shall change their place, their air, their ground, & such like, you pervert their kind; and you shall find that the young growing plant will resemble another kind, both in colour and fashion.

*

It remains that we show how we may fashion mandrakes, those counterfeit kind of mandrakes, which couzeners and cony-catchers carry about, and sell to many instead of true mandrakes. You must get a great root of bryony, or wild nep, and with a sharp instrument engrave in it a man or woman, giving either of them their genitories: and then make holes with a puncheon into those places where the hairs are wont to grow, and put into those holes millet, or some other such thing which may shoot out his roots like the hairs of one's head. And when you have digged a little pit for it in the ground, you must let it lie there, until such time as it shall be covered with a bark, and the roots also be shot forth.

JOHN BAPTISTA PORTA, *Natural Magick*, 1558

The male mandrake hath great broad long smooth leaves of a darke green colour, flat spread upon the ground: among which come up the flowers of a pale whitish colour, standing every one upon a single small and weake foot-stalk of a whitish green colour: in their places grow round apples of a yellowish colour, smooth, soft, and glittering, of a strong smell; in which are contained flat and smooth seeds in the fashion of a little kidney, like those of the thorn-apple. The root is long, thick, whitish, divided many times into two or three parts resembling the legs of a man, with other parts of his body adjoining thereto, as the privy part, as it hath been reported; whereas in truth it is no otherwise than in the roots of carrots, parsnips, and such like, forked or divided into two or more parts, which Nature taketh no account of. There hath been many ridiculous tales brought up of this plant, whether of old wives, or some runnagate surgeons or physick-mongers I know not, (a title bad enough for them) but sure some one or more that sought to make themselves famous and skilful above others, were the first broachers of that error I speak of. They add further, that it is never or very seldom to be found growing naturally but under a gallows, where the matter that hath fallen from the dead body hath given it the shape of a man; and the matter of a woman, the substance of a female plant, with many other such doltish dreams. They fable further and affirm, that he who would take up a plant thereof must tie a dog thereunto to pull it up, which will give a great shriek at the digging up; otherwise if a man should do it, he should surely die in short space after. Besides many fables of loving matters, too full of scurrility to set forth in print, which I forbear to speak of. All which dreams and old wives tales you shall from henceforth cast out of your books and memory; knowing this, that they are all and every part of them false and most untrue: for I my self and my servants also have digged up, planted, and replanted very many, and yet never could either perceive shape of man or woman, but sometimes one straight root, sometimes two, and often six or seven branches coming from the main great root, even as Nature list to bestow upon it, as to other plants. But the idle drones that have little or nothing to do but eat and drink, have bestowed some of their time in carving the roots of bryony, forming them to the shape of men & women: which falsifying practise hath confirmed the error amongst the simple and unlearned people, who have taken them upon their report to be the true mandrakes.

> JOHN GERARD, *The Herball, or General Historie of Plantes*, 1597: a corrective from the sixteenth-century English herbalist John Gerard, who argued that even real mandrakes were a deception

In the West parts of Cornwall, during the winter season, swallows are found sitting in old deep tin-works, and holes of the sea cliffs; but touching their lurking places, *Olaus Magnus* maketh a far stranger report. For he saith, that in the North parts of the world, as summer weareth out, they clap mouth to mouth, wing to wing, and leg in leg, and so after a sweet singing, fall down into certain great lakes or pools amongst the Canes, from whence at the next spring, they receive a new resurrection; and he addeth for proof hereof, that the fishermen, who make holes in the ice, to dip up such fish with their nets, as resort thither for breathing, do sometimes light on these swallows, congealed in clods, of a slimy substance, and that carrying them home to their stoves, the warmth restoreth them to life and flight.

RICHARD CAREW, *The Survey of Cornwall*, 1602

The learned Dr Pierce, physician at Bath, in a letter to the ingenious Mr William Musgrave, Secretary of the Philosophical Society of Oxford, sent us lately an account also of such a Toad found in the centre of a hard lime-stone, laid as a step-stone for passengers in the middle of a cartway between two rills that ran each side it; where a croaking noise being a long time heard, and the parts near searched and nothing found, this stone at length was resolved should be broke, where in a cavity near the middle, a large toad was found as big as a man's fist, which hopped about as briskly, as if it had been bred in a larger room; but for how long time he does not say. But the toad that was found in the most astonishing manner, certainly that ever was heard of, was that at Statfold, if the tradition they have of it there be true, where as the story goes, the steeple being to be taken down to prevent falling, the top-stone of the spire or pinnacle being taken off, was thrown down whole into the church-yard, but breaking in the fall, there appeared a living toad in the centre of it, which (as most of the rest are said to do) died quickly after it was exposed to the air . . .

They are also sometimes met with in this county as closely included in the bodies of firm trees: thus out of a great oak that grew at Lapley of about 6 tunns of timber, brought to Elmhurst, by the right Worshipful Sir Theophilus Biddulph Baronet for the new building the house, there was a great toad sawn forth of the middle of the tree, in a place which when growing, was 12 or 14 foot from the ground; the tree being sound and entire in all parts quite round, saving just where the toad lay, it was black and corrupted, and crumbled away like sawdust. Also at Bently there was another sawn out of a solid tree, in that part of it, that when growing, might be about a yard from the ground; the tree sound underneath next the root, and in all other parts, only where the toad lay, there was a hollow about the bigness of the crown of one's hat, which (as those enclosed in stone) also presently died, as soon as exposed to the air. Now how these animals should come at all to be thus included, in the middle of such entire and solid substances? and when enclosed, how maintained either with breath, or aliment? and how long they may have been presumed, to have continued there? seem questions indeed worthy the consideration of the most profound philosopher; whom that I may honestly provoke to give a better, I shall here offer the reader some account of my own, which though a slender one enough, yet may serve his turn, till he can get a better, and in some measure to evince the probability of the thing.

To come then close to the business, upon presumption that the matter of fact is indisputable; 'tis easy to apprehend how toads creep into the clefts and hollows of rocks and trees (which they always do in August, when they are in a declining condition) to preserve themselves in the winter: where during their rest for about eight months, they grow somewhat bigger, and the clefts or holes of the rocks or trees, as much less; so that at the return of the year (like the fox in the fable) they cannot get out, where they came in, and so are forced to remain where they are, in that solitary condition, as long as they live; the clefts and holes of the rocks and trees in the mean time growing quite up, and enclosing them in an entire and solid case. And thus I suppose these animals may come to be enclosed in the rocks and trees, upon or near the surface of the earth. But how that toad in the tree at Lapley, should come to be thus imprisoned 12 or 14 foot high? is a difficulty yet harder, and that requires yet nicer considerations.

For the solution whereof, we must either suppose that the toad was produced in a hole at that height when the tree was young, of an agreeable dust, brought thither by the wind and a sort of rain as well disposed for the same purpose; like the worms and maggots bred of dust, and the rains that accompany the tornado blasts, and fall in the Maggoti Savanna in Jamaica, by equivocal generation, or else according to the opinion of Cardan, generated of the seed of a toad blown from the top of some mountain; or drawn up by the sun into the clouds, and so discharged thence in a shower, and lodged in the bole of this tree whilst young: whence fearing to leap in the summer, and creeping down low in the dust, usually lodged in the boles of all trees, in the winter, and there keeping its station for a long season; the wood of the tree in a little time might thus grow over it, so that the tree being trimmed up, and a taller body given it, the toad at length thus appeared to be enclosed in the body of the tree at that height.

Nor is it at all improbable that the spawn of toads, or indeed that toads themselves, should be thus drawn up by the sun's heat, since we see what vast quantities of water it supports in those wonderful exhalations they call spouts at sea, in which there are such mighty weights of water, that they overwhelm the best ships, if any thing near them, and disturb the whole sea for a good distance, with the violence of their fall: in these spouts together with the water, the fish many times in the sea thereabout are also lifted up, which sometimes being carried by the winds over land before their fall, has often occasioned the wonderful raining of fish, as it did whitings, at Stansted in the parish of Wrotham in the county of Kent Anno. 1666; and herrings in the south of Scotland, Anno 1684, as his most Sacred Majesty King James the Second most judiciously determined the problem there. Now most certainly the force that could elevate these, may very well be allowed to attract the spawn of toads, or large toads themselves, which being carried by the wind (that bloweth where it listeth) to any place whatever, may also be let fall as well in any the like indeterminate place, and so possibly upon the bole of a tree as well as any where else.

Robert Plot, The Natural History of Staffordshire, 1686

Because you are writing of birds, I will tell you something concerning nightingales imitating men's voice, and repeating their discourses, which is indeed wonderful, and almost incredible, but yet most true, and which I my self heard with these ears, and had experience of, this last Diet at Ratisbone in the year 1546, whilst I lodged there in a common inn at the sign of the Golden Crown. Our host had three nightingales, placed separately, so that each was shut up single by itself in a dark cage. It happened that at that time, being the spring of the year, when those birds are wont to sing indefatigably, and almost incessantly; I was so afflicted with the stone, that I could sleep but very little all night. Then about and after midnight, when there was no noise in the house, but all still, you might have heard strange janglings and emulations of two nightingales, talking one with another, and plainly imitating men's discourses. For my part I was almost astonished with wonder. For they in the night-season, when all was whist and quiet, in conference together produced and repeated whatever they had heard in the day time from the guests talking together, and had thought upon. Those two of them which were most notable, and masters of this art, were scarce ten foot distant one from the other: the third hung more remote, so that I could not so well hear it as I lay in bed. But those two it is wonderful to tell, how they provoked one another, and by answering invited and drew one another to speak. Yet did they not confound their words, talking both together, but rather utter them alternately, or by course. But besides the daily discourse, which they had lately hear of the guests, they did chant out especially two stories one to the other for a long time, even from midnight till morning, so long as there was no noise of men stirring, and that with that native modulation and various inflection of their notes, that no man, unless he were very attentive and heedful, would either have expected from those little creatures, or easily observed. When I asked the host, whether their tongues had been slit, or they taught to speak any thing? He answered no; whether he had observed or did understand what they sung in the night? He likewise denied that. The same said the whole family. But I who could not sleep whole nights together, did greedily and attentively hearken to the birds, greatly admiring their industry and contention. One of the stories was concerning the tapster, or house-knight (as they call them) and his wife, who refused to follow him going into the wars, as he desired her. For the husband endeavoured to persuade his wife, as far as I understand by those birds, in hope of prey, that she would leave her service in that inn, and go along with him into the wars. But she, refusing to follow him, did resolve either to stay at Ratisbone, or go away to Nurenberg. For there had been an earnest and long contention between them about this matter, but (as far as I understood) no body being present besides, and without the privity of the master of the house; and all this dialogue the birds repeated. And if by chance in their wrangling they cast forth any unseemly words, and that ought rather to have been suppressed and kept secret, the birds, as not knowing the difference between modest and immodest, honest and filthy words, did out with them. This dispute and wrangling the birds did often repeat in the night time, as which (as I guessed) did most firmly stick in their memories, and which they had well conned and thought upon. The other was a history or prediction of the war of the Emperor against the Protestants, which was then imminent. For as it were presaging or prophesying they seemed to chant forth the whole business as it afterwards fell out. They did also with that story mingle what had been done before against the Duke of Brunswick. But I suppose those birds had all from the secret conferences of some Noblemen and Captains, which as being in a public inn, might frequently have been had in that place where the birds were kept. These things (as I said) they did in the night, especially after twelve of the clock, when there was a deep silence, repeat. But in the day-time for the most part they were silent, and seemed to do nothing but meditate upon, and revolve with themselves what the guests conferred together about either at table, or else as they walked. I verily had never believed our Pliny writing so many wonderful things concerning these little creatures, had I not my self seen with my eyes, and heard them with my ears uttering such things as I have related. Neither yet can I of a sudden write all, or call to remembrance every particular that I have heard.

FRANCIS WILLUGHBY, *The Ornithology*, 1678

Crossing the river, we left Thanet and came to Sandwich. We went to an inn and stayed there a short time. Then two of us went off to the seashore towards Sandown Castle; the others got ready to explore the town. Under the guidance of Mr Sparkes, a schoolmaster, they walked round the walls and the bastions, now partly ruinous with age, and entered the garden of Caspar Niren, a Belgian, as also the apothecary's shop of Charles Duck, whom we afterwards met in Canterbury. In this place they saw a thing worth remembering, the 'spoils' (if I may so call them) of a serpent fifteen feet long and thicker than an arm. As far as I can hazard a guess, it was a sea serpent; for it was caught by two men among the sandhills near the seashore, after its head had been shattered by small shot discharged from a fowling piece. It was hunting the rabbits, of which there is a vast abundance there, for food; for one or two were extracted from its stomach. These men, as I have said, brought the dead beast to our good friend Charles Duck, were duly rewarded and handed it over; its skin stripped from the flesh and stuffed with hay he still keeps with him as a memento of the event.

> THOMAS JOHNSON, *Descriptio Itineris Plantarum,* 1632; from 'Description of a Journey Undertaken for the Discovery of Plants into the County of Kent, 1632', tr. C. E. Raven *et al.,* 1972. Johnson was a London botanist who inaugurated the practice of taking apprentice herbalists and apothecaries on plant-hunting expeditions. He also produced a delightful edition of Gerard's *Herball* in 1633, famous for its affectionate digs at the author's occasional over-enthusiasm (e.g. of Gerard's account of a peony found apparently wild in Kent, Johnson wrote: 'I have been told that our author himself planted that peony there, and afterwards seemed to find it there by accident: and I do believe it was so, because none before or since has ever seen or heard of it growing wild since in any part of this kingdom').

THE WILL O WHISP OR JACK A LANTHORN

I have often seen these vapours or what ever philosophy may call them but I never wit nessd so remarkable an instance of them as I did last night which has robd me of the little philosophic reasoning which I had—about them I now believe them spirits but I will leave the facts to speak for themselves—There had been a great upstir in the town about the appearance of the ghost of an old woman who had been recently drownd in a well—it was said to appear at the bottom of neighbour Billings close in a large white winding sheet dress & the noise excited the curosity of myself & my neighbour to go out several nights together to see if the ghost woud be kind enough to appear to us & mend our broken faith in its existance but nothing came on our return we saw a light in the north east over eastwell green & I thought at first that it was a bright meoter it presently became larger & seemd like a light in a window it then moved & dancd up & down & then glided onwards as if a man was riding on horsback at full speed with a lanthorn light soon after this we discoverd another rising in the south east on 'dead moor' they was about a furlong asunder at first & as if the other saw it it danced away as if to join it which it soon did & after dancing together a sort of reel as it were—it chaced away to its former station & the other followd it like things at play & after suddenly overtaking it they mingled into one in a moment or else one dissapeard & sunk in the ground we stood wondering & gazing for a while at the odd phenomenon & then left the will o wisp dancing by itself to hunt for a fresh companion as it chose—the night was dusky but not pitch dark & what was rather odd for their appearance the wind blew very briskly it was full west—now these things are gennerally believd to be vapours rising from the foul air from bogs & wet places were they are generaly seen & being as is said lighter then the common air they float about at will—now this is all very well for Mrs Philosophy who is very knowing but how is it if it is a vapour lighter then the air that it coud face the wind which was blowing high & always floated side ways from north to south & back—the wind afected it nothing but I leave all as I find it I have explaind the fact as well as I can—I heard the old alewife at the Exeters arms behind the church (Mrs Nottingham) often say that she has seen from one of her chamber windows as many as fifteen together dancing in & out in a company as if dancing reels & dances on east well moor there is a great many there—I have seen several there myself one night when returning home from Ashton on a courting excursion I saw one as if meeting me I felt very terrified & on getting to a stile I determnd to wait & see if it was a person with a lanthorn or a will o whisp it came on steadily as if on the path way & when it got near me within a poles reach perhaps as I thought it made a sudden stop as if to listen me I then believed it was some one but it blazd out like a whisp of straw & made a crackling noise like straw burning which soon convincd me of its visit the luminous haloo that spread from it was of a mysterious terrific hue & the enlargd size & whiteness of my own hands frit me the rushes appeard to have grown up as large & tall as walebone whips & the bushes seemd to be climbing the sky every thing was extorted out of its own figure & magnified the darkness all round seemd to form a circalar black wall & I fancied that if I took a step forward I shoud fall into a bottomless gulph which seemd yawing all round me so I held fast by the stile post till it darted away when I took to my heels & got home as fast as I coud so much for will o whisps.

JOHN CLARE, *Journals*, 1828; from *The Natural History Writings of John Clare*, ed. Margaret Grainger, 1983

Watching Narrowly

*I*N *1781 Daines Barrington (one of Gilbert White's correspondents in* The Natural History of Selborne*) denied the possibility of high-altitude, and therefore 'invisible', migration, by referring to the infall-ibility of his own eyesight: 'I never lost the sight of a bird myself, but from its horizontal distance, and I doubt very much whether any bird was ever seen to rise to a greater height than perhaps twice that of St Paul's cross.'*

This was a classic Enlightenment argument, anecdotal, dogmatic, and prepared to use the evidence of the senses as an arbiter—just so long as it did not conflict with passionately held beliefs. When the evid-ence itself made qualitative advances, from the revelations of Hooke and Leeuwenhoek's microscopes, for instance, or Gilbert White's sym-pathetic and open-minded fieldwork, the old arrogance and fables began to collapse.

But none of these changes of perspective would have made head-way without a fundamental philosophical shift. The great admission of the time was that nature mattered, that it was worth attention. The trail was blazed for the most functional of reasons by the seventeenth-century herbalists and botanists (they wanted medicinal plants that worked), and later by antiquarians such as Aubrey, who were fasci-nated by historical continuity in nature as much as in the built envir-onment. In the confident mood of the mid-eighteenth century, their ad hoc *investigations developed into more formal kinds of survey—both round the year and across the country—driven by the belief that a comprehensive 'Calendar of Flora' and 'a General Natural History of Great Britain' might eventually unlock the basic patterns and rhythms of natural growth.*

It was the beginning of the science of ecology. Yet it was under-pinned by the Christian philosophy of 'physico-theology' advanced by writers such as Ray and Derham. This was a doctrine that explored and celebrated the idea of a divinely inspired design in nature. To sceptical modern minds it can seem like a collection of truisms: a

natural creation which did not 'work' would simply not exist. But its
commitment to observing and unravelling the real workings of the crea-
tion, rather than taking them on trust, for once put the Church firmly
on the side of scientific advance.

When I was forced, following an illness that affected me both
physically and mentally, to rest from more serious studies, and
to spend my time in riding and walking, I had leisure in the
course of my journeys to contemplate the varied beauty of plants
and the cunning craftsmanship of Nature that was constantly
before my eyes, and had so often been thoughtlessly trodden
underfoot. Once I had become more aware of these wonders, I
ceased to pass them by and treat them as matters unworthy of
my attention.

First I was fascinated and absorbed by the rich spectacle of
the meadows in spring-time; then I was filled with wonder and
delight by the marvellous shape, colour and structure of the individual plants.

While my eyes feasted on these sights, my mind too was stimulated. I became inspired with a passion for Botany, and I conceived a burning desire to become proficient in that study, from
which I promised myself much innocent pleasure to soothe my
solitude.

I searched throughout the University, looking everywhere for
someone to act as my teacher and guide, who would instruct me
and, so to speak, initiate me, so that I would be able to enjoy
the benefit of his advice whenever I needed it. But, to my astonishment, among so many masters of learning and luminaries of
letters I found not a single person who was deeply versed in
Botany, and only one or two who had even a slight acquaintance
with the subject.

Then I began to realise what enormous difficulties would con-
front a novice like myself embarking on this study, and how tor-
tuous would be the windings of the path I would have to tread,
and what a great amount of time and toil would be required to
make even small progress in these studies, whereas with the help
and guidance of a skilled instructor I might have acquired a full
and detailed knowledge of the subject in a minimum of time and
with no great trouble.

What was I to do in this situation? Should I allow the flame of my enthusiasm to be quenched or be diverted to some other field of study? I decided that this must not happen. I was filled with shame and sorrow that at a time when all the other sciences and branches of learning were flourishing and making notable progress in our midst, this precious branch of Natural Philosophy, so useful and indeed essential to common life, should be lying neglected and unheeded. Here was a way by which one of moderate attainments like myself could confer a benefit on the University. Why should not I, endowed with ample leisure, if not with great ability, try to remedy this deficiency so far as my power permitted, and advance the study of Phytology, which had been passed over and neglected by other men?

So I gladly devoted my attention to this study, and derived a rare enjoyment from my work which had the two-fold advantage of being both congenial to myself and beneficial to others.

Spurred on by these considerations, although deprived of any outside help, I had no difficulty in enduring all the toil and tedium, and I untiringly pressed on with my enterprise by seeking out everywhere the different kinds of simples and planting them in my garden; then I made long walks of exploration in the countryside in the vicinity of Cambridge so that I might not appear to close my eyes to the treasures that Nature so freely offers here at home, and, led on by an impatient longing for novelty, eagerly search for strange plants in foreign fields. At last with ceaseless endeavour and untiring effort I overcame the many difficulties, and, after negotiating all the hazards and obstacles in my path, succeeded in reaching my goal.

First of all I had to familiarise myself with the literature of the subject, and then compare the plants that I had found in the countryside with the pictures in the books; then, when I found any similarity between them, I had to study the descriptions more closely. After a time I acquired skill from practice; when I chanced upon some unknown plant, I first considered to what tribe and family it belonged or could be assigned (after I had been occupied for some time with Phytology, I developed a facility in identifying the points of similarity) so I first of all looked for it in the appropriate group, and in this way saved myself a great deal of trouble.

Then I felt a great desire to help the studies of others who might be filled with a love of Botany, and I carefully considered how I could most effectively assist them so that they, perhaps less patient of labour than myself, would not be deterred by the endless succession of difficulties, and falter in their studies. Everybody is always willing to apply himself diligently to what interests him, and—I readily admit it—I was eager to make progress in my studies for my own pleasure.

I have always valued highly the company and friendship of men of ability, and I could think of no way more conducive to arousing their interest so that they might be enticed, if I may use such a word, to share my studies, and no method more effective for attracting their notice and winning their favour than the publication of a Catalogue of all the plants that I had found in the Cambridgeshire countryside, because a common interest in some particular field of study is the strongest bond of association and friendship—for assuredly common interests create a sense of fellow-feeling, and every one is always eager to further his own particular enthusiasm . . .

Certainly no one need fear that such a study would be unproductive and useless, for, if I may quote the words of P. Laurembergius: 'Nothing within the compass of the whole wide world yields a richer pleasure not only to the mind but also to the body, the servant of the mind, than the rich store of plant life, and the copious and varied produce of things growing in the earth.' and a little further on: 'I say that Man receives from plants all the many things which life requires, whether for living simply or in moderation or in luxury. Human frailty has need of food, drink, medicines, clothing, housing, furniture, shipping, the pleasures of the senses and of the mind—all of these needs plants lavish upon us for our use and enjoyment from their store.' as he shows by enumerating each item fully.

I readily admit that, as human affairs are now, such studies do not greatly contribute to the accumulation of wealth or to the winning of the favour of our fellow-men, nevertheless I know of no occupation which is more worthy or more delightful for a free man than to contemplate the beauteous works of Nature and to honour the infinite wisdom and goodness of God the Creator. I do not suggest that any one should deliver and devote himself entirely to these studies, but that he should embrace them within reason and sometimes divert himself with them for his personal pleasure so that he can learn something thoroughly well even in his moments of leisure and not allow any part of his life to be completely empty.

I am quite sure that the pursuit of plants will be a pleasurable occupation for a studious youth for I have known many scholars of every rank in Trinity College for whom this occupation has afforded not only bodily exercise but also mental satisfaction.

I fully realise that not everyone is captivated by the sight of flowers or of the meadows in spring, or if they are captivated, there is something that delights them even more. Some take pleasure in ball-games, others in drinking, gambling, money-making or popularity-hunting, and they show themselves very diligent participants in these activities. I am not writing Phytology for such as these for they are interested in something quite different, but I offer a hundred banquets for the Pythagoreans, dedicated to the true philosophy, whom kindly Nature and Titan have fashioned from finer clay, whose concern is not so much to know what authors think as to gaze with their own eyes on the nature of things and to listen with their own ears to her voice, who prefer to know quality to quantity and usefulness to pretension . . .

> JOHN RAY, preface to *Catalogus Plantarum circa Cantabrigia nascientum*, 1680; tr. A. H. Ewen and C. T. Prime, 1975

[Tobacco] is a herb that goeth creeping up by certain little canes, it hath a sad green colour, he carrieth certain leaves, that the greatness of them may be of the greatness of a good potenge dish, that is in compass round, with a little point, the leaf hath his little sinews, he is small, well near without moisture, the stalk is of the colour of a clear tawny. They say that he doth cast certain clusters, with little grapes, of the greatness of a coriander seed, which is his fruit and doth wax ripe by the month of September: he doth cast out many boughs, the which doth stretch along upon the earth, and if you do put anything near to it, it goeth creeping upon it. The root of the Mechoacan is unsavoury and without biting or any sharpness of taste . . .

One of the marvels of this herb and that which bringeth most admiration is the manner how the Priests of the Indias did use it, which was in this manner: when there was amongst the Indians any manner of business of great importance, in the which the chief gentleman called Casiques or any of the principal people of the country had necessity to consult with their Priests in any business of importance: then they went and propounded their matter to their chief Priest, forthwith in their presence he took certain leaves of the tobacco and cast them into the fire and did receive the smoke of them at his mouth and at his nose with a cane, and in taking of it he fell down upon the ground as a dead man, and remaining so according to the quantity of the smoke that he had taken, when the herb had done his work he did revive and awake, and gave them then answers according to the visions, and illusions which he saw, whilst he was rapt in the same manner, and he did interpret to them as to him seemed best, or as the Devil had counselled him, giving them continually doubtful answers in such sort that howsoever it fell out, they might say that it was the same which was declared and the answer that he made.

In like sort the rest of the Indians for their pastime do take the smoke of the tobacco, to make themselves drunk withall, and to see the visions, and things that represent unto them, that wherein they do delight: and other times they take it to know their business and success, because conformable to that which they have seen, being drunk therewith, even so they judge of their business. And as the devil is a deceiver and hath the knowledge of the virtue of herbs, so he did show the virtue of this herb, that by the means thereof, they might see their imaginations and visions, that he hath represented unto them and by that means deceive them.

> NICHOLAS MONARDES, *Joyfull Newes out of the newe founde worlde*, 1569

To employ some trusty fellow that shall undertake to traverse Cader Idris for a whole day in search of plants, observing punctually these following directions.

Taking a handbasket with him, he must go up as far as Llyn y Cau ere he takes notice of anything: but then he must trace some rivulet of water, as high as he can with safety, putting into his basket ten or a dozen roots of each sort of herb he can discover; but of such as are very small, 15 or 20. He must sometimes stray from the rivulets to the rocks and gather anything he meets with there; which he found not near the rivulets. He must not omit any sort of plant that he sees; excepting the common sort of ffern, Heath and grasse.

Of all the shrubs he meets with: I mean such as Llûs duon, Llûs Cochion, Gruglys, Mwyar y Mynudh, Crâch-helig &c.; let him take up only 3 or 4 roots and those the least. Let him not neglect any sort of Moss: such as Troed y blaudh, Corn y Carw, Mwswgl y ffynodwydh, Mwswgl y Cypreswydh (which creeps abundantly on the grasse towards the Top of the Hill) &c. He must be cautious in picking up the very least thing his eyes can discover; for by the rills of water there are some plants so small as scarce to be seen, which nevertheless are as rare as any.

We don't confine him to the rivulets all day; for when he has searched several of them, he may wander amongst the rocks, pastures &c. on all sides of the Hill; but let him gather nothing that grows lower than a quarter of a mile of the Top; & let him be sure that he'll find the greatest variety by the rivulets of water and other wet place; espec. in Craig dhu above Llyn y Cau; where he must not fail to climb as high as he can with safety. For his further encouragement, when we shall receive these plants, we shall send him by the carrier a farthing for every different kind he has gathered; which if he proves any thing diligent, can not amount to less than 2 shillings.

If there be no possibility of sending to Cader Idris let these directions be observed at Rennogl Fawr.

EDWARD LHWYD, letter to the plant collector David Lloyd, 1682; from *Life and Letters of Edward Lhwyd*, 'Early Science in Oxford', vol. xiv, 1945

LORDS AND LADIES, CUCKOO-PINT. *ARUM MACULATUM* L.

Local names. ADAM AND EVE, Som, Leic, Lincs, Yks; ADDER'S FOOD, Som; ADDER'S MEAT (cf. German *Schlangenbeeren*, 'snake berries'), Corn, Dev, Som; ADDER'S TONGUE, Corn, Som; ANGELS AND DEVILS (cf. German *Engelcher und Deiwelche*), Som; ARON, Scot.

BABE-IN-THE-CRADLE (cf. German *Kinneken in der Wieg*), Som; BLOODY FINGERS, Hants; BLOODY MAN'S FINGER (i.e. devil's finger), Som, Worc; BOBBIN AND JOAN, N'hants; BOBBIN JOAN, Corn; BULLOCKS, Som; BULLS AND COWS, Som, N'hants, Lincs, Lancs, Yks; BULLS AND WHEYS ('whey' or 'quey', a heifer), Yks, West; BULLS, Dor.

CALVES' FOOT (translation of the old botanists' name *Pes vituli*), Som; COCKY BABY, I o W; COWS AND CALVES, Dev, Dor, Som, Wilts, Glos, Bucks, N'hants, War, Worc, Shrop, Notts, Lincs, Yks, Lakes; COWS AND KIES, Yks; COW'S PARSNIP, Som; CUCKOO COCK, Ess; CUCKOO-FLOWER, N'hants; CUCKOO-PINT (i.e. pintle, penis. See also *Orchis mascula*), Suss, E Ang, N'hants, Leic; CUCKOO-POINT, Yks.

DEAD MAN'S FINGERS, Worc; DEVILS AND ANGELS (see above), Dor, Som; DEVILS, LADIES AND GENTLEMEN, Denb; DEVIL'S MEN AND WOMEN, Shrop; DOG BOBBINS, N'hants; DOG COCKS, Wilts; DOG'S DIBBLE, Dev; DOG'S SPEAR, DOG'S TASSEL, Som; FAIRIES, Som; FLY-CATCHER, Wilts; FROG'S MEAT, Dor.

GENTLEMEN AND LADIES, Oxf; GENTLEMEN'S AND LADIES' FINGERS, GENTLEMAN'S FINGER, Wilts; GREAT DRAGON (see below), Suss; HOBBLE-GOBBLES, Kent; JACK-IN-THE-BOX, Som, Bucks, N Ire; JACK-IN-THE-GREEN, Som; JACK-IN-THE-PULPIT, Corn, Som, Lincs; KINGS AND QUEENS, Som, Lincs, Dur; KITTY-COME-DOWN-THE-LANE-JUMP-UP-AND-KISS-ME, Kent; KNIGHTS AND LADIES, Som.

LADIES AND GENTLEMEN, Som, Wilts, Kent, N'hants, Shrop; LADIES' LORDS, Kent; LADY'S FINGER, Wilts, Glos, Kent; LADY'S KEYS, Kent; LADY'S SLIPPER, Wilts; LADY'S SMOCK, Dor, Som, Hants; LAMB-IN-A-PULPIT, Dev, Wilts; LAMB'S LAKENS (i.e. 'toys'), N'hants, N'thum, N Eng; LILY, Wilts; LILY GRASS, Suss.

LONG PURPLES, War; LORDS AND LADIES, general from Cornwall to Lakes and Yks; LORDS' AND LADIES' FINGERS, War.

MANDRAKE, Yks; MAN-IN-THE-PULPIT, MEN AND WOMEN, Som; MOLL OF THE WOODS (cf. *Anemone nemorosa*), War; NIGHTINGALE, Ess; OLD MAN'S PULPIT, Som; OXBERRY, Worc; PARSON AND CLERK, Dev, Som; PARSON-IN-HIS-SMOCK, Lincs; PARSON IN THE PULPIT, Dev, Dor, Som, Ches, Yks; PARSON'S BILLYCOCK (i.e. pintle, see *Orchis mascula* and Shakespeare, *King Lear*, III. iv. 74–5), PREACHER-IN-THE-PULPIT, Som; PRIESTIES, Lancs; PRIEST-IN-THE-PULPIT, Som; PRIEST'S PILLY (i.e. pintle), West; PRIEST'S PINTLE (i.e. penis), Derb, Lincs, Dur, Cumb; POISON-FINGERS, Dor; POISON-ROOT, Wilts; POKERS, Som.

RAM'S HORN, Suss; RAMSON, Cumb; RED-HOT POKER, Som; SCHOOL-MASTER, Suss; SILLY LOVERS, Som; SMALL DRAGON (see below), Suss; SNAKE'S FOOD, Dev, Som; SNAKE'S MEAT, Dev; SNAKE'S VICTUALS, Wilts, Glos; SOLDIERS, Som; SOLDIERS AND ANGELS, Dev; SOLDIERS AND SAILORS, Som; STALLIONS, STALLIONS AND MARES, Lincs, Yks; STANDING GUSSES, SUCKY CALVES, SWEETHEARTS, Som; TOAD'S MEAT, Corn.

WAKE ROBIN, Corn, Dor, Suss, Berks, War, Worc, Ches, Yks, Scot, N Ire; WHITE AND RED, Dor; WILD LILY, Dev.

> GEOFFREY GRIGSON, *The Englishman's Flora*, 1958: a collection of vernacular names for the cuckoo-pint, many of

which date back to the Middle Ages, and demonstrate the perennial fascination of ordinary people with the resemblances and allusions of plant anatomy

OF CUCKOO PINT, OR WAKE-ROBIN.
THE DESCRIPTION.

Arum or cuckoo pint hath great, large, smooth, shining, sharp pointed leaves, bespotted here and there with blackish spots, mixed with some blueness: among which riseth up a stalk nine inches long, bespeckled in many places with certain purple spots. It beareth also a certain long hose or hood, in proportion like the ear of an hare: in the middle of which hood cometh forth a pestle or clapper of a dark murrie or pale purple colour: which being past, there succeedeth in place thereof a bunch or cluster of berries in manner of a bunch of grapes, green at the first, but after they be ripe of a yellowish red like coral, and full of pith, with some thready additaments annexed thereto.

OF SOW-BREAD.
THE DESCRIPTION.

The first being the common kind of sowbread, called in shops *Panis porcinus*, and *Arthanita*, hath many green and round leaves like unto Asarabacca, saving that the upper part of the leaves are mixed here and there confusedly with white spots, and under the leaves next the ground of a purple colour: among which rise up little stems like unto the stalks of violets, bearing at the top small purple flowers, which turn themselves backward (being full blown) like a Turks cap, or tulepan, of a small scent or savour, or none at all: which being past there succeed little round knops or heads which contain slender brown seeds: these knops are wrapped after a few days in the small stalks, as thread about a bottom, where it remaineth so defended from the injury of winter close upon the ground, covered also with the green leaves aforesaid, by which means it is kept from the frost, even from the time of his seeding, which is in September until June: at which time the leaves do fade away, the stalks and seed remaining bare and naked, whereby it enjoyeth the sun (whereof it was long deprived) the sooner to bring them unto maturity: the root is round like a turnip, black without and white within, with many small strings annexed thereto . . .

THE DANGER.

It is not good for women with child to touch or take this herb, or to come near unto it, or stride over the same where it groweth for the natural attractive virtue therein contained is such, that without controversy they that attempt it in manner abovesaid, shall be delivered before their time: which danger and inconvenience to avoid, I have (about the place where it groweth in my garden) fastened sticks in the ground, and some other sticks I have fastened also cross-ways over them, lest any woman should by lamentable experiment find my words to be true, by their stepping over the same.

> JOHN GERARD, *The Herball*, Thomas Johnson edition, 1633. Johnson commented on Gerard's sowbread notes: 'I judge our Author something too womanish in this, that is, led more by vain opinion than by any reason or experience, to confirm this his assertion, which frequent experience shows to be vain and frivolous, especially for the touching, striding over, or coming near this herb.'

HELLEBORINE ANGUSTIFOLIA
NARROW-LEAVED WILD WHITE HELLEBORE

Epipactis palustris L.
Grantz

The flower of this plant consists of five petals, which wind round a small body like that of a fly or insect, whose white inner part is marked with purple lines or streaks; the outer part (if you examine the three outer petals) is a dull purple with some green and white marking.

The remaining two inner petals have the same colour on both sides, inside as well as outside. The stylus like a body of the insect hidden within these petals has a yellowish head with a white body veined with purple streaks; the white stomach is joined to the chest with a thin band over a groove, and is thus easily separable. The overhanging labellum or lip is full, white and as if fringed on the edge. Tiny pointed single leaves grow from each separate stalk of the flowers, and on the top of the stalk are three or four smaller leaves in a bundle forming an apex. The upper parts of the stalk where the flowers are attached to it, are sprinkled with a kind of flour as are the calyces of the flower. The root creeps under the ground and so propagates itself easily. I have not observed this plant growing anywhere except in marshy and watery localities, for which reason I wonder why it should be described as 'montana' by C.B.

RAPUM SYLVESTRE
WILD RAPE OR TURNEP

Brassica napus L.
Sown in the Isle of Ely.

Caterpillars born on brassica have taught us that a close relationship exists between these stocks as the leaves of rape are eaten no less greedily than those of brassica although they scorn many other plants that we have offered them as food.

Let us follow up the history of these caterpillars since I have never read or seen anything like it; let me describe it even if it is too prolix to suit the plan of my book; at least we shall be worthy of the reader's pardon.

The caterpillar which eats brassica, if you examine it, is of average size, and is clothed with whitish short hairs at intervals which are at no point close together. The colour of its body is black, yellow and blue in a variegated pattern. There are three yellow lines running the length of the body, one down the middle of the back and the others on the two sides; the blue and black colour lies interspersed between these lines, the black in spots and the blue spread out. These black spots project from the body and hairs grow out from the centre of each. The colour of the head is composed of the same three colours and the head is also covered with similar hairs. The feet are sixteen in number arranged in three groups, a first group of six, the middle group of eight (as in the majority of caterpillars) and two in the last group. Such is the outward appearance of the caterpillar.

I shut up ten or so of these in a wooden box at the end of August 1658. They fed for a few days and fixed themselves to the sides or lid of the box. Seven of them proved to be viviparous or vermiparous; from their backs and sides very many, from thirty to sixty apiece wormlike animalcules broke out; they were white, glabrous, footless and under the microscope transparent. As soon as they were born, they began to spin silken cocoons, finished them in a couple of hours, and in early October came out as flies, black all over with reddish legs and long antennae, and about the size of a small ant. The three or four caterpillars which did not produce maggots changed into angular and humped

chrysalids which came out in April as white butterflies. The viviparous caterpillars died a few days after the birth without any
metamorphosis. The fact is also worthy of notice that if you
observe the said caterpillars at the beginning of autumn, you will
find more of them viviparous; if you observe them when Autumn
is almost ended, you will find more transformed into chrysalids.

SAMBUCUS
COMMON ELDER

Sambucus nigra L.

N.I. A plum tree grafted into Elder, although I have tried it and
it does not easily take, bears purgative fruits. *Cam. hort.*
2. This tree produces a peculiar kind of fungus which the Jews
call 'the ear' on account of its similarity in appearance, and from
this 'signature' they deduce that it is good for ears.

And at this point one can note in passing the foolishness of
the chemists who chatter and boast so loudly of the signatures
of plants. We have paid close attention to the matter and are
moved to assert that the signatures are not indications of natural qualities and powers impressed on plants by nature. 1. Of the
plants specifically said to be appropriate to a particular part of
the body or to a disease far the greater number have no signature as it is easy to demonstrate in the heart, the chest, the skull
and the liver. 2. Different parts of the same plant such as the
leaves, roots, flowers and seeds, have signatures not merely different but contradictory. 3. Many plants resemble natural or
artificial objects for which they have no affinity as Orchid flowers
look like spiders, frogs, bees or butterflies. Between these (as I
may note in passing) so great a similarity exists that no one could
fail to recognize it immediately nor could he persuade himself
that it had ever come about by chance; it is clear proof of inten-
tion on the part of nature. The seed cases of Shepherds's Purse
resemble little bags, Thlaspi small trays, Antirrhinum vulgaris a
calf's head; Pisum cordatum has the shape of a heart impressed
on it as have the leaves of some species of Medick; Tragopogon
a goat's beard; the root of the earth nut a mouse, Iris and Gladiolus
a sword, the foliage of fenugreek horns: to which objects they
cannot be related. 4. Parts of some plants represent parts of the
body with which they violently disagree. Thus the fruit of Anacar-
ditum represents a heart and is nevertheless poisonous; the juice
of the spurges is like milk but no one is such a imbecile as to
give it to nursing mothers; the flesh of Mespilus is like human

excrement and similar in colour, but is not suitable as an aperi-
ent. The fungus ignarius growing on the trunks of trees recalls
the lungs in shape but it is nonetheless harmful and dangerous
to them. It would be endless to enumerate every example. 5. The
same parts of the body are affected by different and often contradictory diseases which require remedies differing in character.
6. In different plants, parts of the same appearance and shape
possess different and diverse properties; the bulbous roots of some,
like Narcissus, promote vomiting while those like the root of tulip
form a very pleasant food. 7. Neither are the number of signatures so great nor the signatures they bear so obvious and plain
to anybody that they suggest a pointer or deliberate plan on the
part of Nature. There is such a vast number of plants that even
if they had come into existence altogether and by chance, any
ingenious and imaginative person could have found as many signatures as are known today.

> JOHN RAY, *Catalogus Plantarum circa Cantabrigia nascientum*,
> 1680; tr. A. H. Ewen and C. T. Prime, 1975

The principal misfortune of botany is, that from its very birth it
has been looked upon merely as a part of medicine. This was
the reason why everybody was employed in finding or supposing
virtues in plants, whilst the knowledge of plants themselves was
totally neglected: for how could the same man make such long
and repeated excursions as so extensive a study demands; and at
the same time apply himself to the sedentary labours of the laboratory, and attendance upon the sick; which are the only methods of ascertaining the nature of vegetable substances, and their
effects upon the human body? This false idea of botany, for a
long time, almost confined the study of it to medicinal plants, and
reduced the vegetable chain to a small number of interrupted links.
Even these were very ill studies, because the substance only was
attended to, and not the organisation. How indeed could persons be much interested in the organical structure of a substance,
of which they had no other idea but as a thing to be pounded
in a mortar! Plants were searched for, only to find remedies; it
was simples, not vegetables, that they looked after. This was very
right, it will be said. Maybe so. Hence nevertheless it follows,
that, if men were ever so well acquainted with remedies, they were
very ignorant of plants; and this is all that I have here advanced.

*

LETTER TO A LADY
ON THE FRUCTIFICATION AND LILIACEOUS PLANTS

Dated the 22nd of August 1771

I think your idea of amusing the vivacity of your daughter a little, and exercising her attention upon such agreeable and varied
objects as plants, is excellent: though I should not have ventured
to play the pedant so far as to propose it of myself. Since however it comes from you, I approve it with all my heart, and will
even assist you in it; convinced that, at all times of life, the study
of nature abates the taste for frivolous amusements, prevents the
tumult of the passions, and provides the mind with a nourishment which is salutary, by filling it with an object most worthy
of its contemplations.

You have begun with teaching your daughter the names of the
common plants which you have about you; this was the very thing
you should have done. The few plants which she knows by sight
are so many points of comparison for her to extend her knowledge: but they are not sufficient. You desire to have a little catalogue of the most common plants, with the marks by which they
may be known. I find some difficulty in doing this for you: that
is, in giving you these marks or characters in writing, after a manner that is clear, and at the same time not diffuse. This seems
impossible without using the language peculiar to the subject;
and the terms of that language form a vocabulary apart, which
you cannot understand unless it be previously explained to you.

Besides, merely to be acquainted with plants by sight, and to
know only their names, cannot but be too insipid a study for a
genius like yours; and it may be presumed that your daughter
would not be long amused with it. I propose that you should
have some preliminary notions of the vegetable structure or organisation of plants, in order that you may get some real information, though you were to take only a few steps, into the most
beautiful and the richest of the three kingdoms of nature. We
have nothing therefore to do yet with the nomenclature, which
is but the knowledge of a herbarist. I have always thought it possible to be a very great botanist without knowing so much as one
plant by name; and, without wishing to make your daughter a
very great botanist, I think nevertheless that it will always be useful to her to learn how to see, whatever she looks at, well. Do
not however be terrified at the undertaking: you will soon know
that it is not a great one. There is nothing either complicated or

difficult in what I have to propose to you. Nothing is required but to have patience to begin with the beginning. After that, you may go on no farther than you choose.

We are now getting towards the latter season, and those plants which are the most simple in their structure are already past. Besides, I expect you will take some time to make your observations a little regularly. However, in the meanwhile, till spring puts you in a situation to begin and follow the order of nature, I am going to give you a few words of the vocabulary to get by heart.

A perfect plant is composed of a root, of a stem with its branches, of leaves, flower, and fruit, (for in botany, by fruit, in herbs as well as in trees, we understand the whole fabric of the seed.) You know the whole of this already, at least enough to understand the term; but there is a principal part which requires an examination more at large; I mean the *fructification*, that is, the *flower* and the *fruit*. Let us begin with the flower, which comes first. In this part nature has enclosed the summary of her work; by this she perpetuates it, and this also is commonly the most brilliant of all parts of the vegetable, and always least liable to variations.

> Jean-Jacques Rousseau, *Letters on the Elements of Botany: Addressed to a Lady*, 1785; tr. Thomas Martyn, 1785

Elms. I never did see an elm that grew spontaneously in a wood, as oaks, ashes, beeches, &c.; which consideration made me reflect that they are exotic; but by whom were they brought into this island? Not by the Saxons; for upon enquiry I am informed that there are none in Saxony, nor in Denmark, nor yet in France, spontaneous; but in Italy they are natural; e.g. in Lombardy, &c. Wherefore I am induced to believe that they were brought hither out of Italy by the Romans, who were cultivators of their colonies. The Saxons understood not nor cared for such improvements, nor had hardly leisure if they would.

Anno 1687 I travelled from London as far as the Bishopric of Durham. From Stamford to the bishopric I saw not one elm on the roads, whereas from London to Stamford they are in every hedge almost. In Yorkshire is plenty of trees, which they call elms; but they are witch-hazels, as we call them in Wilts (in some counties wych-elms). I acquainted Mr. Jo. Ray of this, and he told me when he travelled into the north he minded it not, being chiefly intent on herbs; but he writes the contrary to what I do here: but it is matter of fact, and therefore easily to be proved.

In the Villare Anglicanum are a great many towns, called *Ash*ton, *Willough*by, &c, but not above three or four *Elme*-tons.

In the common at Urshfont was a mighty elm, which was blown down by the great wind when Ol. Cromwell died. I saw it as it lay along, and I could but just look over it.

*

Since the writing this of elms, Edmund Wyld, Esq. of Houghton Conquest in Bedfordshire, R.S.S. assures me that in Bedfordshire, in several woods, e.g. about Wotton, &c. that elms do grow naturally, as ashes, beeches &c.; but query, what kind of elm it is?

*

Beeches. None in Wilts except at Groveley. I have a conceit that long time ago Salisbury plains might have woods of them, but that they cut them down as an incumbrance to the ground, which would turn to better profit by pasture and arable. The Chiltern of Buckinghamshire is much of the like soil; and there the nearness of Bucks to London, with the benefit of the Thames, makes their woods a very profitable commodity.

*

About the middle of Groveley Forest was a fair wood of oaks, which was called Sturton's Hat. It appeared a good deal higher than the rest of the forest (which was most coppice wood), and was seen over all Salisbury plains. In the middle of this hat of trees (it resembled a hat) there was a tall beech, which overtopped all the rest. The hat was cut down by Philip II. Earl of Pembroke, 1654; and Thomas, Earl of Pembroke, disafforested it, anno 1684.

*

Yew trees naturally grow in chalky countries. The greatest plenty of them, as I believe, in the west of England is at Nunton Ewetrees. Between Knighton Ashes and Downton the ground produces them all along; but at Nunton they are a wood. At Ewridge, in the parish of Colern, in North Wilts (a stone brash and a free stone), they also grow indifferently plentiful; and in the parish of Kington St. Michael I remember three or four in the stone brash and red earth.

When I learnt my accidents, 1633, at Yatton Keynel, there was a fair and spreading yew-tree in the churchyard, as was common heretofore. The boys took much delight in its shade, and it furnished them with their scoops and nutcrackers. The clerk lopped it to make money of it to some bowyer or fletcher, and that lopping killed it: the dead trunk remains there still.

*

Box, a parish so called in North Wilts, near Bath, in which parish is our famous freestone quarry of Haselbery: in all probability took its name from the box-trees which grew there naturally, but now worn out.

Not far off on Cotswold in Gloucestershire is a village called Boxwell, where is a great wood of it, which once in . . . years Mr. Huntley fells, and sells to the comb-makers in London. At Boxley in Kent, and at Boxhill in Surrey, both chalky soils, are great box woods, to which the comb-makers resort.

*

Holly is indifferently common in Malmesbury hundred, and also on the borders of the New Forest: it seems to indicate pit-coal. In Wardour Park are holly trees that bear yellow berries. I think I have seen the like in Cranborne Chase.

*

Hazel. We have two sorts of them. In the south part, and particularly Cranbourn Chase, the hazels are white and tough; with which there are made the best hurdles of England. The nuts of the chase are of great note, and are sold yearly beyond sea. They sell them at Woodbery Hill Fair, &c.; and the price of them is the price of a bushel of wheat. The hazel trees in North Wilts are red, and not so tough, more brittle.

*

In the old hedges which are the bounds between the lands of Priory St. Marie, juxta Kington St. Michael, and the west field, which belonged to the Lord Abbot of Glastonbury, are yet remaining a great number of berberry-trees, which I suppose the nuns made use of for confections, and they taught the young ladies that were educated there such arts. In those days there were not schools for young ladies as now, but they were educated at religious houses.

> JOHN AUBREY, 'Of Plants', *Memoires of Naturall Remarques in the County of Wiltshire*, 1685

OF THE RHINOCEROS

We are now to discourse of the second wonder in nature, namely of a beast every way admirable, both for the outward shape, quantity and greatness, and also for the inward courage, disposition, and mildness. For as the elephant was the first wonder, of whom we have already discoursed, so this beast next unto the Elephant filleth up the number, being every way as admirable as he, if he does not exceed him, except in quantity or height of stature; and being now come to the story of this beast, I am heartily sorry, that so strange an outside, yielding no doubt through the omnipotent power of the creator, an answerable inside, and infinite testimonies of worthy and memorable virtues comprised in it, should through the ignorance of men, lie unfolded and obscured before the readers' eyes: for he that shall but see our stories of the apes, of the dogs, of the Mice, and of other small beasts, and consider how large a treatise we have collected together out of many writers, for the illustration of their natures and vulgar conditions, he cannot choose but expect some rare and strange matters, as much unknown to his mind about the story of this rhinoceros, as the outward shape and picture of him, appeareth rare and admirable in his eyes: differing in every part from all other beasts, from the top of his nose to the tip of his tail, the ears and eyes excepted, which are like bears. But gentle Reader as thou art a man, so thou must consider since Adam went out of Paradise, there was never any that was able perfectly to describe the universal conditions of all sorts of beasts, and it hath been the council of the almighty himself, for the instruction of man, concerning his fall and natural weakness, to keep him from the knowledge of many divine things, and also human, which is of birds and beasts, fishes and fowl, that so he might learn, the difference betwixt his generations, and his degeneration, and consider how great a loss unto him was his fall in Paradise.

> CONRAD GESNER, *Historiae Animalium*, 1551: from Edward Topsell's translation of 1607, which gave a warm and

personal touch to Gesner's rather dry text (and humanized
this rare confession of zoological ignorance)

THE GREAT GREY SHRIKE

The bill is black and moderately short and hooked at the tip: it
is the stoutest and strongest of all, so much so that once it wounded my hand even though I was protected by a double glove; it
very quickly breaks and crushes the bones and skulls of birds. . . .
It has short wings and flies as if it were jumping up and down.
It lives on beetles, butterflies and the larger insects, but not only
on these: like a hawk it lives on birds. It kills Goldcrests and
Finches and (as I once observed) Thrushes. Bird-catchers even
report that it sometimes slays certain woodland Pies and drives
away Crows. It does not fly down the birds that it kills and strike
them with its claws like Hawks, but ambushes them and attacks
them and (as I have often noted) aims at the throat, and squeezes
and breaks the skull with its beak. It fractures and crushes the
bones and then devours them, and when it is hungry crams into
its throat such gobbets of flesh as the narrow gape can contain.
Moreover unlike other birds when it has abundance of prey it
stores up some of it against future shortage. It impales and hangs
big flies and insects that it has caught on the thorns and spines
of bushes. It is of all birds the most easily tamed; and when used
to the hand is fed on meat. If this is dry or bloodless, it requires
drink. I have seen it in England not more than twice; in Germany
very frequently.

> WILLIAM TURNER, *Avium praecipuarum*, 1544; tr. Charles E.
> Raven, 1947

In the year 1675, about half-way through September (being busy
with studying air, when I had much compressed it by means of
water), I discovered living creatures in rain, which had stood but
a few days in a new tub, that was painted blue within. This observation provoked me to investigate this water more narrowly; and
especially because these little animals were, to my eye, more than
ten thousand times smaller than the animalcule which Swammerdam
has portrayed, and called by the name of Water-flea, or Water-
louse, which you can see alive and moving in water with the bare
eye.

Of the first sort that I discovered in the said water, I saw, after divers observations, that the bodies consisted of 5, 6, 7, or 8 very clear globules, but without being able to discern any membrane or skin that held these globules together, or in which they were inclosed. When these animalcules bestirred 'emselves, they sometimes stuck out two little horns, which were continually moved, after the fashion of a horse's ears. The part between these little horns was flat, their body else being roundish, save only that it ran somewhat to a point at the hind end; at which pointed end it had a tail, near four times as long as the whole body, and looking as thick, when viewed through my microscope, as a spider's web. At the end of this tail there was a pellet, of the bigness of one of the globules of the body; and this tail I could not perceive to be used by them for their movements in very clear water. These little animals were the most wretched creatures that I have ever seen; for when, with the pellet, they did but hit on any particles or little filaments (of which there are many in water, especially if it hath but stood some days), they stuck intangled in them; and then pulled their body out into an oval, and did struggle, by stretching themselves, to get their tail loose; whereby their whole body then sprang back towards the pellet of the tail, and their tails then coiled up serpent-wise, after the fashion of a copper or iron wire that, having been wound close about a round stick, and then taken off, kept all its windings. This motion, of stretching out and pulling together the tail, continued; and I have seen several hundred animalcules, caught fast by one another in a few filaments, lying within the compass of a coarse grain of sand.

I also discovered a second sort of animalcules, whose figure was an oval; and I imagined that their head was placed at the pointed end. These were a little bit bigger than the animalcules first mentioned. Their belly is flat, provided with divers incredibly thin little feet, or little legs, which were moved very nimbly, and which I was able to discover only after sundry great efforts, and wherewith they brought off incredibly quick motions. The upper part of their body was round, and furnished inside with 8, 10, or 12 globules: otherwise these animalcules were very clear. These little animals would change their body into a perfect round, but mostly when they came to lie high and dry. Their body was also very yielding: for if they so much as brushed against a tiny filament, their body bent in, which bend also presently sprang out again; just as if you stuck your finger into a bladder full of water, and then, on removing the finger, the inpitting went away. Yet the greatest marvel was when I brought any of the animalcules on a dry place, for I then saw them change themselves at last into a round, and then the upper part of the body rose up pyramid-like, with a point jutting out in the middle; and after having thus lain moving with their feet for a little while, they burst asunder, and the globules and a watery humour flowed away on all sides, without my being able to discern even the least sign of any skin wherein these globules and the liquid had, to all appearance, been inclosed; and at such times I could discern more globules than when they were alive. This bursting asunder I figure to myself to happen thus: imagine, for example, that you have a sheep's bladder filled with shot, peas, and water; then, if you were to dash it apieces on the ground, the shot, peas, and water would scatter themselves all over the place.

Furthermore, I discovered a third sort of little animals, that were about twice as long as broad, and to my eye quite eight times smaller than the animalcules first mentioned: and I imagined, although they were so small, that I could yet make out their little legs, or little fins. Their motion was very quick, both roundabout and in a straight line.

The fourth sort of animalcules, which I also saw a-moving, were so small, that for my part I can't assign any figure to 'em. These little animals were more than a thousand times less than the eye of a full-grown louse (for I judge the diameter of the louse's eye to be more than ten times as long as that of the said creature), and they surpassed in quickness the animalcules already spoken of. I have divers times seen them standing still, as 'twere, in one spot, and twirling themselves round with a swiftness such as you see in a whip-top a-spinning before your eyes; and then again they had a circular motion, the circumference whereof was no bigger than that of a small sand-grain; and anon they would go straight ahead, or their course would be crooked. . . .

ANTON VAN LEEUWENHOEK, 'Concerning Little Animals Observed . . . in Rain- Well- Sea- and Snow-Water', *Philosophical Transactions of the Royal Society*, 1677. Leeuwenhoek was a minor public official in the Dutch city of Delft, but spent most of his time developing the microscope. His proof of the existence of creatures too small to be seen by the naked eye in that most universal of mediums, water, was a revelation, and had a profound effect on contemporary natural philosophy. His papers to the Royal Society were originally contributed as letters written in Old Dutch (he had no English) and were translated by a sympathetic but unknown member of the Society

The last remark I shall make about the terraqueous globe in general is, the great variety of kinds, or tribes, as well as prodigious number of individuals of each various tribe, there is of all creatures. There are so many beasts, so many birds, so many insects, so many reptiles, so many trees, so many plants upon the land; so many fishes, sea-plants, and other creatures in the waters; so many minerals, metals, and fossils in the subterraneous regions; so many species of these genera, so many individuals of those species, that there is nothing wanting to the use of man, or any other creature of this lower world. If every age doth change its food, its way of clothing, its way of building; if every age hath its variety of diseases; nay, if man, or any other animal, was minded to change these things every day, still the creation would not be exhausted, still nothing would be wanting for food, nothing for physic, nothing for building and habitation, nothing for cleanliness and refreshment, yea, even for recreation and pleasure. But the munificence of the Creator is such, that there is abundantly enough to supply the wants, the conveniences, yea, almost the extravagancies of all the creatures, in all places, all ages, and upon all occasions.

And this may serve to answer an objection against the excellency of, and wisdom showed in, the creation; namely, what need of so many creatures? Particularly of so many insects, so many plants, and so many other things? And especially of some of them, that are so far from being useful, that they are very noxious; some by their ferity, and others by their poisonous nature, etc.

To which I might answer, that in great variety, the greater art is seen; that the fierce, poisonous, and noxious creatures serve as rods and scourges to chastise us, as means to excite our wisdom, care and industry, with more to the same purpose. But these things have been fully urged by others; and it is sufficient to say, that this great variety is a most wise provision for all the uses of the world in all ages, and all places. Some for food, some for physic, some for habitation, some for utensils, some for tools and instruments of work, and some for recreation and pleasure, either to man or to some of the inferior creatures themselves; even for which inferior creatures the liberal Creator hath provided all things necessary, or any ways conducing to their happy, comfortable living in this world, as well as for man.

*

There is a great deal of geometrical neatness and nicety, in the sinuous motion of snakes, and other serpents. For the assisting in which action, the annular seales under their body are very remarkable, lying cross the belly, contrary to what those in the back, and the rest of the body do; also as the edges of the foremost scales lie over the edges of their following scales, from head to tail; so those edges run out a little beyond, or over their following scales; so as that when each scale is drawn back, or set a little upright, by its muscle, the outer edge thereof, or foot it may be called, is raised also a little from the body, to lay hold on the earth, and so promote and facilitate the serpent's motion. This is what may be easily seen in the slough, or belly of the serpent-kind. But there is another admirable piece of mechanism, that my antipathy to those animals hath prevented my prying into; and that is, that every scale hath a distinct muscle, one end of which is tacked to the middle of its scale; the other, to the upper edge of its following scale.

*

The wise author of nature, having denied feet and claws to enable snails to creep and climb, hath made them amends in a way more commodious for their state of life, by the broad skin along each side of the belly, and the undulating motion observable there. By this latter it is they creep; by the former, assisted with the glutinous slime emitted from the snail's body, they adhere firmly and securely to all kinds of superficies, partly by the tenacity of their slime, and partly by the pressure of the atmosphere.

*

The motive parts, and motion of caterpillars, are useful, not only to their progression and conveyance from place to place; but also to their more certain, easy, and commodious gathering of food: for having feet before and behind, they are not only enabled to go by a kind of steps made by their fore and hind parts; but also to climb up vegetables, and to reach from their boughs and stalks

for food at a distance; for which services their feet are very nice-
ly made both before and behind. Behind, they have broad palms
for sticking to, and these beset almost round with small sharp
nails, to hold and grasp what they are upon; before, their feet
are sharp and hooked, to draw leaves, etc, to them, and to hold
the fore-part of the body, whilst the hinder-parts are brought up
thereto. But nothing is more remarkable in these reptiles, than
that these parts and motions are only temporary, and incompar-
ably adapted only to their present nympha-state; whereas in their
aurelia-state, they have neither feet nor motion, only a little in
their hinder-parts: and in the mature-state, they have the parts
and motion of a flying insect, made for flight.

> WILLIAM DERHAM, 'The Great Variety and Quantity of all
> Things . . .', *Physico-theology*, 1713, 1754

It is a common remark among people, who are but little acquainted with the works of Nature, that the sensitive plants approach
very near to the animal kingdom, at least to those plant-like sea
productions which have lately been proved to be real animals;
because these plants, when irritated ever so little, show a kind of
sensation or motion, by contracting their leaves together, particularly in that genus called *Mimosa* by Linnaeus, and remarkably
in that species of it called *Mimosa pudica*, or what we call the
humble plant, where not only the leaves contract on the touch,
but the young joints bend down: besides this genus, there is also
an *Oxalis*, or wood sorrel of the East Indies, that has a sensitive
quality of contracting its pennated leaves on the least touch.

This extraordinary operation of Nature, that surprises us so
much, has often been attempted to be explained by many ingenious men; and accounts have been published, but without that
satisfactory clearness to the public, which is always expected from
the sensible investigators of Nature. It seems to be a secret that
still lies hid, and possible will lie hid from the strictest investigation of human philosophy.

Indeed the leaves of the sensitive plants, that we have been
hitherto acquainted with, are so minute and tender, that they
cannot be so well dissected. But for the satisfaction of the curious in this way, we have fortunately received from Pennsylvania,
very lately, a new genus of plants, quite different from any thing
heretofore described, whose leaves are succulent, and large enough

for dissection, and formed in a manner not only new and surprising, but likewise very entertaining; having at the end of each
leaf two lobes, or lips, in the shape of the eye-lids, an inch broad,
furnished with a row of stiff hairs on the margin of each, so that
upon the introducing a straw or pin between them, they contract
themselves, and grasp it quite close. This plant being an inhab-
itant of a warmer country than this, the gardeners observe that
it is most active in a hot-bed, though it seems to thrive very well
in this country in the open air. The following account is what
we have been able to collect of the history of this curious plant:

About three years ago that diligent and indefatigable botanist,
Mr. John Battram, an honest sober Quaker of Philadelphia, sent
a dried specimen of this extraordinary plant in flower to the worthy Peter Collinson, Esq. of Mill Hill, F.R.S. the lately deceased,
much-lamented friend of all botanists, by the Indian name, either
Cherokee or Catabaw, but which I cannot now recollect, of Tippity-
wichit, which he said he had collected in the swamps beyond the
Blue Mountains. At the request of Mr. Collinson, the ingenious
Doctor Solander, now on his voyage to the South Seas, in search
of the rarer productions of Nature, dissected this plant before
some of his friends; and from the beautiful appearance of its milk-
white flowers, and the elegance of its leaves, thought it well
deserved one of the names of the Goddess of Beauty, and therefore called it *Dionæa*.

As this name was generally approved of, and so well adapted
by that eminent botanist, I shall only add a specific name to
distinguish it from others of this genus, that may possibly be
discovered hereafter. From the structure then and particular
moving quality of its leaves when irritated, I shall call it *Dionæa
muscipula*, which may be construed into English, with humble
submission both to critics and foreign commentators, either
Venus's Flytrap, or Venus's Mousetrap.

I have looked into the Index of the intelligent Mr. Miller's
Gardener's Dictionary for a precedent, and find that there are
plants which have formerly been called after that Goddess, as
Venus's Looking-Glass, and Venus's Navel-Wort, and both adopted
by him.

I presume then that the name of Venus's Flytrap, as it seems
most adapted to its powers, may be admitted to be the most eli-
gible trivial name, especially as I think myself warranted to do it from
ocular demonstration of this surprising faculty of its entrapping

little animals, such as earwigs, spiders, and flies, where they are either squeezed to death or remain imprisoned till they die.

JAMES ELLIS, 'Venus's Mousetrap', St James's Chronicle, 1768

The whole earth would be overwhelmed with carcases, and stinking bodies, if some animals did not delight to feed upon them. Therefore when an animal dies, bears, wolves, foxes, ravens, &c. do not lose a moment till they have taken all away. But if a horse e.g. dies near the public road, you will find him, after a few days, swollen, burst, and at last filled with innumerable grubs of carnivorous flies, by which he is entirely consumed, and removed out of the way, that he may not become a nuisance to passengers by his poisonous stench.

When the carcases of fishes are driven upon the shore, the voracious kinds, such as the thornback, the hound fish, the conger eel, &c. gather about and eat them. But because the flux, and reflux soon change the state of the sea, they themselves are often detained in pits, and become a prey to the wild beasts, that frequent the shores. Thus the earth is not only kept clean from the putrefaction of carcases, but at the same time by the economy of nature the necessaries of life are provided for many animals. In the like manner many insects at once promote their own good, and that of other animals. Thus gnats lay their eggs in stagnant, putrid and stinking waters, and the grubs that arise from these eggs clear away all the putrefaction; and this will easily appear, if any one will make the experiment by filling two vessels with putrid water, leaving the grubs in one, and taking them all out of the other. For then he will soon find the water, that is full of grubs, pure and without any stench, while the water that has no grubs will continue stinking.

The beetle kind in summer extract all moist and glutinous matter out of the dung of cattle, so that it becomes like dust, and is spread by the wind over the ground. Were it not for this, the vegetables that lie under the dung, would be so far from thriving, that all that spot would be rendered barren.

As the excrements of dogs is of so filthy and septic a nature, that no insect will touch them, and therefore they cannot be dispersed by that means, care is taken that these animals should exonerate upon stones, trunks of trees, or some high place, that vegetables may not be hurt by them.

Cats bury their dung. Nothing is so mean, nothing so little, in which the wonderful order, and wise disposition of nature does not shine forth.

> Isaac Biberg, *The Oeconomy of Nature*, 1749. Biberg was a perceptive Swedish naturalist and pioneer ecologist, and a colleague of Carl Linnaeus. His long paper *The Oeconomy of Nature* was included in translation in Benjamin Stillingfleet's influential *Miscellaneous Tracts Relating to Natural History*, 1759

Words cannot express the joy that the sun brings to all living things . . . Yes, Love comes even to the plants. Males and females, even the hermaphrodites, hold their nuptials (which is the subject that I now propose to discuss), showing by their sexual organs which are males, which females, which hermaphrodites . . . The actual petals of a flower contribute nothing to generation, serving only as the bridal bed which the great Creator has so gloriously prepared, adorned with such precious bed-curtains, and perfumed with so many sweet scents in order that the bridegroom and bride may therein celebrate their nuptials with the greater solemnity. When the bed has thus been made ready, then is the time for the bridegroom to embrace his beloved bride and surrender himself to her.

*

I noticed that she was blood-red before flowering, but that as soon as she blooms her petals become flesh-coloured. I doubt whether any artist could rival these charms in a portrait of a young girl, or adorn her cheeks with such beauties as are here and to which no cosmetics have lent their aid. As I looked at her I was reminded of Andromeda as described by the poets, and the more I thought about her the more affinity she seemed to have with the plant; indeed, had Ovid set out to describe the plant mystically [*mystice*] he could not have caught a better likeness. . . .

Her beauty is preserved only so long as she remains a virgin (as often happens with women also)—i.e. until she is fertilized, which will not now be long as she is a bride. She is anchored far out in the water, set always on a little tuft in the marsh and fast tied as if on a rock in the midst of the sea. The water comes up to her knees, above her roots; and she is always surrounded by poisonous dragons and beasts—i.e. evil toads and frogs—which drench her with water when they mate in the spring. She stands and bows her head in grief. Then her little clusters of flowers with their rosy cheeks droop and grow ever paler and paler. . . .

> CARL LINNAEUS, *Journals*, tr. Wilfrid Blunt, 1971. Linnaeus devised the current system of biological nomenclature, allotting every species one family or generic name, and one specific. It was a classification based on the sexual apparatus of the plants, and Linnaeus's innocent attempts to popularize it by using human marriage as an analogy produced frequent moments of bawdy bedroom farce (e.g. 'Polyandria: Twenty males or more in the same bed with the female') that were often shocking to contemporary readers

THAWING MONTH

From the first melting of the snow to the floating of ice down rivers.

Mar III

 xix. Eves drop towards the noontide sun.
   Sallow, round leaved, flower-buds, 449. 15 Salix caprea, open.
 xx. Snow melts against walls.
   Lark begins to sing.
 xxii. Water flows by the walls.
 xxv. Roads very dirty and full of water.

April IV

 i. Horse dung melts the ice.
   Moss, upright fir. Lycopodium selago, 106. sheds its dust.
 iii. Stones are loosened from the ice.
 vi. Hills begin to appear, the snow being melted.
   Serpents come out of their holes.
   Spider, water, frisks about. The fly creeps forth.
   Game, black, 53. Tetrao tetrix.
   Lapwing, 110. Tringa vanellus, returns.
 vii. Butterfly, nettle, Papilio urticae, appears in abundance.

Some people, says Pliny, think the appearance of the butterfly the surest sign of spring, on account of the delicacy of the animal.

> Duck, tame, 145. Anas Boschas, sits.
> Wild duck returns.

    x. An inundation of snow water.

        Swan, 37. Anas cygnus, and Daker-hen, 58.8. Rallus crex,
           by their appearance proclaim the spring.

        Rivers are unbound, and ice floats down.

N.B. The river at Upsal, for 70 years, has never been frozen
beyond the 19th of April, according to the observation of O.
Celsius, sen.

        Pike, 112. Esox lucius, spawns. This fish gives over spawn-
           ing when the frog begins.

    xi. Snow water soaks into the earth.

        Subterraneous places are inundated.

        Frog comes forth.

        Winter shelters ought to be removed from garden plants,
           that they may not be too much drawn up.

        Hot-beds for melons should be sown.

            Alexander Berger, *The Calendar of Flora*, 1755. This cal-
            endar, made in Uppsala, Sweden, in 1755, was the inspira-
            tion for a comparable journal written up in the village of
            Stratton in Norfolk by Benjamin Stillingfleet. This in turn
            encouraged Daines Barrington, in 1767, to devise a printed
            diary blank—'The Naturalist's Journal'—which was subse-
            quently used by Gilbert White to record his daily weather,
            gardening, and natural history observations

My Brother Tho & I went down with a spade to examine into
the nature of those animals that make that chearful shrill cry all
the summer months in many parts of the south of England. We
found them to be of the Cricket-kind, with wings & ornamented Cases over them, like the House kind. But tho' they have long
legs behind with large brawny thighs, like Grasshoppers, for leaping; it is remarkable that when they were dug-out of their holes
they shewed no manner of activity, but crawled along in a very
shiftless manner, so as easily to be taken. We found it difficult
not to squeese them to death in breaking the Ground: & out of
one so bruised I took a multitude of eggs, which were long, of
a yellow Colour, & covered with a very tough skin.

    It was easy to discover the male from the female; the former
of which is of a black shining Colour, with a golden stripe across
it's shoulders something like that of the Humble bee: the latter
was more dusky, & distinguished by a long terebra at it's tail,

which probably may be the instrument with which it may deposit it's eggs in Crannies, & safe receptacles. It is very likely that the males only make that shrilling noise; which they may do out of rivalry, & emulation during their breeding time; as is the Case with many animals. They are solitary Insects living singly in Holes by themselves; & will fight fiercely when they meet, as I found by some which I put into an hole in a dry wall, where I should be glad to have them encrease on account of their pleasing summer sound. For tho' they had express'd distress by being taken out of their knowledge; yet the first that had got possession of the chink seized an other with a vast pair of serrated fangs so as to make it cry-out. With these strong, tooth'd Malae (like the sheers of lobster's claws) they must terebrate their curious regular Holes; as they have no feet suited for digging like the mole-cricket. I could but wonder, that when taken in hand, they never offer'd to bite, tho' furnish'd with such formidable weapons. They are remarkably shy, & cautious, never stirring but a few inches from the mouth of their holes, & retiring backward nimbly into them, & stopping short in their song by that time you come within several yards of their caverns: from whence I conclude they may be a very desirable food to some animals, perhaps several kinds of birds. They cry all night as well as day during part of the month of May, June, & July in fine weather; & may in the still part of the night be heard to a considerable distance; abounding most in sand-banks on the sides of heaths, especially in Surrey, & Sussex: but these that I caught were in a steep, rocky pasture-field facing to the afternoon sun.

GILBERT WHITE, *Garden Kalendar*, 20 May 1761: White's first long natural history note

A few house-martins begin to appear about the sixteenth of April; usually some few days later than the swallow. For some time after they appear the *hirundines* in general pay no attention to the business of nidification, but play and sport about either to recruit from the fatigue of their journey, if they do migrate at all, or else that their blood may recover its true tone and texture after it has been so long benumbed by the severities of winter. About the middle of May, if the weather be fine, the martin begins to think in earnest of providing a mansion for its family. The crust or shell of this nest seems to be formed of such dirt or loam as comes most readily to hand, and is tempered and wrought to-gether with little bits of broken straws to render it tough and tenacious. As this bird often builds against a perpendicular wall without any projecting ledge under, it requires its utmost efforts to get the first foundation firmly fixed, so that it may safely carry the superstructure. On this occasion the bird not only clings with its claws, but partly supports itself by strongly inclining its tail against the wall, making that a fulcrum; and thus steadied it works and plasters the materials into the face of the brick or stone. But then, that this work may not, while it is soft and green, pull itself down by its own weight, the provident architect has prudence and forbearance enough not to advance her work too fast; but by building only in the morning, and by dedicating the rest of the day to food and amusement, gives it sufficient time to dry and harden. About half an inch seems to be a sufficient layer for a day. Thus careful workmen when they build mud-walls (informed at first perhaps by this little bird) raise but a moderate layer at a time, and then desist; lest the work should become top-heavy, and so be ruined by its own weight. By this method in about ten or twelve days is formed an hemispheric nest with a small aper-ture towards the top, strong, compact, and warm; and perfectly fitted for all the purposes for which it was intended. But then nothing is more common than for the house-sparrow, as soon as the shell is finished, to seize on it as its own, to eject the owner, and to line it after its own manner.

After so much labour is bestowed in erecting a mansion, as nature seldom works in vain, martins will breed on for several years together in the same nest, where it happens to be well shel-tered and secure from the injuries of weather. The shell or crust of the nest is a sort of rustic work full of knobs and protuber-ances on the outside: nor is the inside of those that I have exam-ined smoothed with any exactness at all; but is rendered soft and warm, and fit for incubation, by a lining of small straws, grass-es, and feathers; and sometimes by a bed of moss interwoven with wool. In this nest they tread, or engender, frequently dur-ing the time of building; and the hen lays from three to five white eggs.

At first when the young are hatched, and are in a naked and helpless condition, the parent birds, with tender assiduity, carry out what comes away from their young. Was it not for this affec-tionate cleanliness the nestlings would soon be burnt up, and destroyed in so deep and hollow a nest, by their own caustic excrement. In the quadruped creation the same neat precaution is made use of; particularly among dogs and cats, where the dams lick away what proceeds from their young. But in birds there seems to be a particular provision, that the dung of nestlings is enveloped into a tough kind of jelly, and therefore is the easier conveyed off without soiling or daubing. Yet, as nature is cleanly in all her ways, the young perform this office for themselves in a little time by thrusting their tails out at the aperture of their nest. As the young of small birds presently arrive at their ἡλικία or full growth, they soon become impatient of confinement, and sit all day with their heads out at the orifice, where the dams, by clinging to the nest, supply them with food from morning to night. For a time the young are fed on the wing by their parents; but the feat is done by so quick and almost imperceptible a sleight, that a person must have attended very exactly to their motions before he would be able to perceive it. As soon as the young are able to shift for themselves, the dams immediately turn their thoughts to the business of a second brood: while the first flight, shaken off and rejected by their nurses, congregate in great flocks, and are the birds that are seen clustering and hovering on sunny mornings and evenings round towers and steeples, and on the roofs of churches and houses. These congregations usually begin to take place about the first week in August; and therefore we may conclude that by that time the first flight is pretty well over. The young of this species do not quit their abodes all together; but the more forward birds get abroad some days before the rest. These approaching the eaves of buildings, and playing about before them, make people think that several old ones attend one nest. They are often capricious in fixing on a nesting place, beginning many edifices, and leaving them unfinished; but when once a nest is completed in a sheltered place, it serves for several seasons. Those which breed in a ready finished house get the start in hatching of those that build new by ten days or a fortnight. These industrious artificers are at their labours in the long days before four in the morning: when they fix their materials they plaster them on with their chins, moving their heads with a quick vibratory motion. They dip and wash as they fly sometimes in very hot weather, but not so frequently as swallows. It has been observed that martins usually build to a north-east or north-west aspect, that the heat of the sun may not crack and destroy their nests: but instances are also remembered where they bred for many years in vast abundance in an hot stifled inn-yard, against a wall facing to the south.

Birds in general are wise in their choice of situation: but in this neighbourhood every summer is seen a strong proof to the contrary at an house without eaves in an exposed district, where some martins build year by year in the corners of the windows. But, as the corners of these windows (which face to the south-east and south-west) are too shallow, the nests are washed down every hard rain; and yet these birds drudge on to no purpose from summer to summer, without changing their aspect or house. It is a piteous sight to see them labouring when half their nest is washed away and bringing dirt . . . '*generis lapsi sarcire ruinas.*' Thus is instinct a most wonderful unequal faculty; in some instances so much above reason, in other respects so far below it! Martins love to frequent towns, especially if there are great lakes and rivers at hand; nay, they even affect the close air of London. And I have not only seen them nesting in the Borough, but even in the Strand and Fleet-street; but then it was obvious from the dinginess of their aspect that their feathers partook of the filth of that sooty atmosphere. Martins are by far the least agile of the four species; their wings and tails are short, and there-fore they are not capable of such surprising turns and quick and glancing evolutions as the swallow. Accordingly they make use of a placid easy motion in a middle region of the air, seldom mounting to any great height, and never sweeping long together over the surface of the ground or water. They do not wander far for food, but affect sheltered districts, over some lake, or under some hanging wood, or in some hollow vale, especially in windy weather. They breed the latest of all the swallow kind: in 1772 they had nestlings on to October the twenty-first, and are never without unfledged young as late as Michaelmas.

As the summer declines the congregating flocks increase in numbers daily by the constant accession of the second broods; till at last they swarm in myriads upon myriads round the villages on the Thames, darkening the face of the sky as they fre-quent the aits of that river, where they roost. They retire, the bulk of them I mean, in vast flocks together about the beginning of October: but have appeared of late years in a considerable flight in this neighbourhood, for one day or two, as late as November the third and sixth, after they were supposed to have been gone for more than a fortnight. They therefore withdraw with us the latest of any species. Unless these birds are very short-lived indeed, or unless they do not return to the district where they are bred, they must undergo vast devastations somehow, and somewhere; for the birds that return yearly bear no manner of proportion to the birds that retire.

House-martins are distinguished from their congeners by having their legs covered with soft downy feathers down to their toes. They are no songsters; but twitter in a pretty inward soft manner in their nests. During the time of breeding they are often greatly molested with fleas.

> GILBERT WHITE, 'An Account of the House-Martin, or Martlet', *Philosophical Transactions of the Royal Society*, 1774: later included in *The Natural History of Selborne*, 1789

A Circumstance struck me the other day in *your way*, it seem'd a novelty to me, but it may be usual & constant, for ought I know. We have great Numbers of Jackdaws, which get under our Tilings. Out of my Study window I have the long Roof of the Deanery before me, and it was new to me that during this whole Month of Decr, as far as it is pass'd, the Jackdaws keep in Pairs. I observed on the Ridge Tiles that tho' a Number were there at a time, yet for the most part they left little spaces, & the Pairs were discernable & separated from the rest; they were likewise in different Pairs on the Declivity of the Roof. It wants much of Valentine's day, but the world is in a Hurry *to secure it's rights*.

> JOHN MULSO, letter to Gilbert White, 15 Dec. 1788. The Revd John Mulso was White's closest friend and most loyal admirer, and his letters provide some of the few surviving glimpses of White's private life. Although Mulso was no naturalist, the publication of *The Natural History of Selborne* moved him to pen this delightful note to White about the courtship of jackdaws

In the year 1774, being much indisposed both in mind and body, incapable of diverting myself either with company or books, and yet in a condition that made some diversion necessary, I was glad of any thing that would engage my attention without fatiguing it. The children of a neighbour of mine had a leveret given them for a plaything; it was at that time about three months old.

Understanding better how to tease the poor creature than to feed it, and soon becoming weary of their charge, they readily consented that their father, who saw it pining and growing leaner every day, should offer it to my acceptance. I was willing enough to take the prisoner under my protection, perceiving that in the management of such an animal, and in the attempt to tame it, I should find just that sort of employment which my case required. It was soon known among the neighbours that I was pleased with the present; and the consequence was, that in a short time I had as many leverets offered to me as would have stocked a paddock. I undertook the care of three, which it is necessary that I should here distinguish by the names I gave them, Puss, Tiney, and Bess. Notwithstanding the two feminine appellatives, I must inform you that they were all males. Immediately commencing carpenter, I built them houses to sleep in; each had a separate apartment so contrived that their ordure would pass thro' the bottom of it; an earthen pan placed under each received whatsoever fell, which being duly emptied and washed, they were thus kept perfectly sweet and clean. In the daytime they had the range of a hall, and at night retired each to his own bed, never intruding into that of another.

Puss grew presently familiar, would leap into my lap, raise himself upon his hinder feet, and bite the hair from my temples. He would suffer me to take him up and to carry him about in my arms, and has more then once fallen asleep upon my knee. He was ill three days, during which time I nursed him, kept him apart from his fellows that they might not molest him (for, like many other wild animals, they persecute one of their own species that is sick), and, by constant care and trying him with a variety of herbs, restored him to perfect health. No creature could be more grateful than my patient after his recovery; a sentiment which he most significantly expressed, by licking my hand, first the back of it, then the palm, then every finger separately, then between all the fingers, as if anxious to leave no part of it unsaluted, a ceremony which he never performed but once again upon a similar occasion. Finding him extremely tractable, I made it my custom to carry him always after breakfast into the garden, where he hid himself generally under the leaves of a cucumber vine, sleeping or chewing the cud till evening; in the leaves also of that vine he found a favourite repast. I had not long habituated him to this taste of liberty, before he began to be impatient for the return of the time when he might enjoy it. He would invite me to the garden by drumming upon my knee, and by a look of such expression as it was not possible to misinterpret. If this rhetoric did not immediately succeed, he would take the skirt of my coat between his teeth, and pull at it with all his force. Thus Puss might be said to be perfectly tamed, the shyness of his nature was done away, and on the whole it was visible, by many symptoms which I have not room to enumerate, that he was happier in human society than when shut up with his natural companions.

Not so Tiney. Upon him the kindest treatment had not the least effect. He too was sick, and in his sickness had an equal share of my attention; but if, after his recovery I took the liberty to stroke him, he would grunt, strike with his fore feet, spring forward and bite. He was, however, very entertaining in his way, even his surliness was matter of mirth, and in his play he preserved such an air of gravity, and performed his feats with such ·a solemnity of manner, that in him too I had an agreeable companion.

Bess, who died soon after he was full grown, and whose death was occasioned by his being turned into his box which had been washed, while it was yet damp, was a hare of great humour and drollery. Puss was tamed by gentle usage; Tiney was not to be tamed at all; and Bess had a courage and confidence that made him tame from the beginning. I always admitted them into the parlour after supper, when the carpet affording their feet a firm hold, they would frisk and bound and play a thousand gambols, in which, Bess, being remarkably strong and fearless, was always superior to the rest, and proved himself the Vestris of the party. One evening the cat being in the room had the hardiness to pat Bess upon the cheek, an indignity which he resented by drumming upon her back with such violence, that the cat was happy to escape from under his paws and hide herself.

You observe, Sir, that I describe these animals as having each a character of his own. Such they were in fact, and their countenances were so expressive of that character, that, when I looked only on the face of either, I immediately knew which it was. It is said, that a shepherd, however numerous his flock, soon becomes so familiar with their features, that he can by that indication only distinguish each from all the rest, and yet to a common observer the difference is hardly perceptible. I doubt not that the same discrimination in the cast of countenances would be discoverable in hares, and am persuaded that among a thousand of them no two could be found exactly similar; a circumstance little suspected by those who have not had opportunity to observe it. These creatures have a singular sagacity in discovering the minutest alteration that is made in the place to which they are accustomed, and instantly apply their nose to the examination of a new object. A small hole being burnt in the carpet, it was mended with a patch, and that patch in a moment underwent the strictest scrutiny. They seem too to be very much directed by the smell in the choice of their favourites; so some persons, though they saw them daily, they could never be reconciled, and would even scream when they attempted to touch them; but a miller coming in, engaged their affections at once; his powdered coat had charms that were irresistible. You will not wonder, Sir, that my intimate acquaintance with these specimens of the kind has taught me to hold the sportsman's amusement in abhorrence; he little knows what amiable creatures he persecutes, of what gratitude they are capable, how cheerful they are in their spirits, what enjoyment they have of life, and that, impressed as they seem with a peculiar dread of man, it is only because man gives them peculiar cause for it.

That I may not be tedious, I will just give you a short summary of those articles of diet that suit them best, and then retire to make room for some more important correspondent.

I take it to be a general opinion that they graze, but it is an erroneous one, at least grass is not their staple; they seem rather to use it medicinally, soon quitting it for leaves of almost any kind. Sowthistle, dent-de-lion, and lettuce are their favourite vegetables, especially the last. I discovered by accident that fine white sand is in great estimation with them; I suppose as a digestive. It happened that I was cleaning a bird-cage while the hares were with me; I placed a pot filled with such sand upon the floor, to which being at once directed by a strong instinct, they devoured it voraciously; since that time I have generally taken care to see them well supplied with it. They account green corn a delicacy, both blade and stalk, but the ear they seldom eat; straw of any kind, especially wheat-straw, is another of their dainties; they will feed greedily upon oats, but if furnished with clean straw never want them; it serves them also for a bed, and, if shaken up daily, will keep sweet and dry for a considerable time. They do not indeed require aromatic herbs, but will eat a small quantity of them with great relish, and are particularly fond of the plant called musk; they seem to resemble sheep in this, that, if their pasture be too succulent, they are very subject to the rot; to prevent which, I always made bread their principal nourishment, and, filling a pan with it cut into small squares, placed it every evening in their chambers, for they feed only at evening and in the night; during the winter, when vegetables are not to be got, I mingled this mess of bread which shreds of carrot, adding to it the rind of apples cut extremely thin; for tho' they are fond of the paring, the apple itself disgusts them. These, however, not being a sufficient substitute for the juice of summer herbs, they must at this time be supplied with water; but so placed, that they cannot overset it into their beds. I must not omit that occasionally they are much pleased with twigs of hawthorn and of the common briar, eating even the very wood when it is of considerable thickness.

Bess, I have said, died young; Tiney lived to be nine years old, and died at last, I have reason to think, of some hurt in his loins by a fall. Puss is still living, and has just completed his tenth year, discovering no signs of decay nor even of age, except that he is grown more discreet and less frolicksome than he was. I cannot conclude, Sir, without informing you that I have lately introduced a dog to his acquaintance, a spaniel that had never seen a hare to a hare that had never seen a spaniel. I did it with great caution, but there was no real need of it. Puss discovered no token of fear, nor Marquis the least symptom of hostility. There is therefore, it should seem, no natural antipathy between dog and hare, but the pursuit of the one occasions the flight of the other, and the dog pursues because he is trained to it: they eat bread at the same time out of the same hand, and are in all respects sociable and friendly. Yours, &c.   W.C.

P S. I should do complete justice to my subject, did I not add, that they have no ill scent belonging to them, that they are indefatigably nice in keeping themselves clean, for which purpose nature has furnished them with a brush under each foot; and that they are never infested by any vermin.

WILLIAM COWPER, *The Gentleman's Magazine*, June 1784

The Romantics

On the surface, there is something paradoxical about the Romantic era being simultaneously a time of a heightened sense of self, and of a heightened sympathy with the 'otherness' of the natural world. Yet even in the early Picturesque writers there is a clear link between these two aspects of Romanticism. Gilpin's writing on forest trees, for example (like Richard Payne Knight's epic poem, The Landscape), celebrates the individuality of living things, and the sense of wildness that Wordsworth commemorated in his line 'A wilderness is rich in liberty'.

The diaries of the period are also rarely wholly subjective or wholly objective. More often they are records of an experience, a moment, some transaction between nature and writer. Only in the worst Romantic writers did they degenerate into what came to be called 'the Pathetic Fallacy', the illusion that human emotions are reflected in the natural world. In Coleridge and Richard Jefferies, especially, what one sees is more a sense of resonance between the two worlds.

I cannot well conceive what your author means by the Common Laws of Nature; but if you desire my opinion how many laws Nature hath, and what they are; I say Nature hath but one law, which is a wise law, *viz.* to keep infinite matter in order, and to keep so much peace, as not to disturb the foundation of her government: for though Nature's actions are various, and so many times opposite, which would seem to make wars between several parts, yet those active parts, being united into one infinite body, cannot break Nature's general peace; for that which Man names war, sickness, sleep, death, and the like, are but various particular actions of the only matter; not, as your author imagines, in a confusion, like bullets, or such like things juggled together in a man's hat, but very orderly and methodical: and the playing motions of nature are the actions of art, but her serious actions are the actions of production, generation and transformation in several kinds, sorts and particulars of her creatures, as also the action of ruling and governing these her several active parts. Concerning the preeminence and prerogative of Man, whom your author call *The flower and chief of all the products of nature upon this Globe of the earth*; I answer, that Man cannot well be judged of himself, because he is a party, and so may be partial; but if we observe well, we shall find that the elemental creatures are as excellent as Man, and as able to be a friend or foe to Man, as Man to them, and so the rest of all creatures; so that I cannot perceive more abilities in Man than in the rest of natural creatures; for though he can build a stately house, yet he cannot make a honey-comb; and though he can plant a slip, yet he cannot make a tree; though he can make a sword, or knife, yet he cannot make the metal: And as Man makes use of other creatures, so other creatures make use of Man, as far as he is good for any thing: but Man is not so useful to his neighbour or fellow-creatures, as his neighbour or fellow-creatures to him, being not so profitable for use, as apt to make spoil.

MARGARET CAVENDISH, *Philosophical Letters*, 1664. Margaret Cavendish, Duchess of Newcastle, was a philosopher, amateur scientist, and writer with a strikingly idiosyncratic voice. She rejected the whole human-centred tradition of her age, and was arguing for the inherent 'wisdom' of the natural world nearly two centuries before the Romantics proper

Since there is nothing which seems more fatally to threaten a weakening, if not a dissolution of the strength of this famous and flourishing nation, than the sensible and notorious decay of her wooden walls, when either through time, negligence, or other accident, the present navy shall be worn out and impaired; it has been a very worthy and seasonable advertisement in the Honourable the principal Officers and Commissioners, what they have lately suggested to this illustrious Society, for the timely prevention and redress of this intolerable defect. For it has not been the late increase of shipping alone, the multiplication of glass-works, iron-furnaces, and the like, from whence this impolitic diminution of our timber has proceeded; but from the disproportionate spreading of tillage, caused through that prodigious havoc made by such as lately professing themselves against root and branch (either to be reimbursed their holy purchases, or for some other sordid respect) were tempted, not only to fell and cut down, but utterly to extirpate, demolish, and raze, as it were, all those many goodly woods, and forests, which our more prudent ancestors left standing, for the ornament, and service of their country. And this devastation is now become so epidemical, that unless some favourable expedient offer itself, and a way be seriously, and speedily resolved upon, for a future store, one of the most glorious, and considerable bulwarks of this nation, will, within a short time, be totally wanting to it.

To attend now a spontaneous supply of these decayed materials (which is the vulgar, and natural way) would cost (besides the enclosure) some entire ages repose of the plough: therefore, the most expeditious, and obvious method, would doubtless be, one of these two ways, sowing, or planting. But, first, it will be requisite to agree upon the species; as what trees are likely to be of greatest use, and the fittest to be cultivated; and then, to consider of the manner how it may best be effected. Truly, the waste, and destruction of our woods, has been so universal, that I conceive nothing less than an universal plantation of all the sorts of trees will supply, and well encounter the defect; and therefore, I shall here adventure to speak something in general of them all; though I chiefly insist upon the propagation of such only as seem to be the most wanting, and serviceable to the end proposed.

> JOHN EVELYN, Introduction to *Sylva, or a Discourse of Forest Trees*, 1678

Besides these requisites of beauty in a tree, there are other things of an adventitious kind, which often add great beauty to it. And here I cannot help lamenting the capricious nature of picturesque ideas. In many instances they run counter to utility; and in nothing more than in the adventitious beauties ascribed to trees. Many of these are derived from the injuries the tree receives, or the diseases, to which it is subject. Mr. Lawson, a naturalist of the last age, thus enumerates them. 'How many forests, and woods, says he, have we, wherein you shall have, for one lively, thriving tree, four, nay sometimes twenty-four, evil thriving, rotten, and dying trees: what rottenness! what hollowness! what dead arms! withered tops! curtailed trunks! what loads of mosses! drooping boughs, and dying branches, shall you see every where.'

Now all these maladies, which our distressed naturalist bemoans with so much feeling, are often capital sources of picturesque beauty, both in the wild scenes of nature, and in artificial landscape.

What is more beautiful, for instance, on a rugged foreground, than an old tree with a hollow trunk? or with a dead arm, a drooping bough, or a dying branch? all which phrases I apprehend are nearly synonymous.

From the withered top also great use, and beauty may result in the composition of landscape; when we wish to break the regularity of some continued line; which we would not entirely hide.

By the curtailed trunk I suppose Mr. Lawson means a tree, whose principal stem has been shattered by winds, or some other accident; while the lower part of it is left in vigour. This is also a beautiful circumstance; and it's application equally useful in landscape. The withered top just breaks the lines of an eminence: the curtailed trunk discovers the whole: while the lateral branches, which are vigorous, and healthy in both, hide any part of the lower landscape, which wanting variety, is better veiled.

For the use, and beauty of the withered top, and curtailed trunk, we need only appeal to the works of Salvator Rosa, in many of which we find them of great use. Salvator had often occasion for an object on his foregrounds, as large as the trunk of a tree; when the whole tree together in it's full state of grandeur, would have been an incumbrance to him. A young tree, or a bush, might probably have served his purpose with regard to composition; but such dwarfs, and striplings could not have preserved the dignity of his subject, like the ruins of a noble tree. These splendid remnants of decaying grandeur speak to the imagination in a style of eloquence, which the stripling cannot reach: they record the history of some storm, some blast of lightening, or other great event, which transfers it's grand ideas to the landscape; and in the representation of elevated subjects assists the sublime.

WILLIAM GILPIN, *Remarks on Forest Scenery*, 1791

BROWN with Sorrel are whole fields lying fallow.
BLUE of the brightest are sloping fields covered with *Echium*, surpassing in splendour anything that can be imagined.

YELLOW and brightly gleaming are the fields of *Chrysanthemum*,
former ploughed fields of *Hypericum* and sand-fields of *Stoechas
citrina*.

RED as blood are often whole slopes of *Viscaria*.

WHITE as snow are sand-fields of the sweet-smelling *Dianthus*.

DAPPLED are the waysides with *Echium, Cichorium, Anchusa* and
Malva . . .

CARL LINNAEUS, *Skanska resa*, 1751

On this cold day about noon a bat was flying round Gracious
street pond, & dipping down & sipping the water, like swallows,
as it flew: all the while the wind was very sharp, & the boys were
standing on the ice!

GILBERT WHITE, *Naturalist's Journal*, 1 Feb. 1785

*The following journal and notebook entries were written during the
period Samuel Coleridge, William Wordsworth, and his sister Dorothy
were living together at Grasmere in the English Lake District.*

In the morning, I read Mr Knight's *Landscape*. After tea we rowed
down to Loughrigg Fell, visited the white foxglove, gathered wild
strawberries, and walked up to view Rydale. We lay a long time
looking at the lake: the shores all embrowned with the scorching
sun. The ferns were turning yellow, that is, here and there one
was quite turned. We walked round by Benson's wood home.
The lake was now most still, and reflected the beautiful yellow
and blue and purple and grey colours of the sky. We heard a
strange sound in the Bainriggs wood, as we were floating on the
water; it *seemed* in the wood, but it must have been above it, for
presently we saw a raven very high above us. It called out, and
the dome of the sky seemed to echo the sound. It called again
and again as it flew onwards, and the mountains gave back the
sound, seeming as if from their center; a musical bell-like answering to the bird's hoarse voice. We heard both the call of the bird,
and the echo, after we could see him no longer.

DOROTHY WORDSWORTH, 27 July 1800

The beards of thistle and dandelions flying above the lonely mountains like life, and I saw them thro' the trees skimming the lake
like swallows.

SAMUEL COLERIDGE, 1 Sept. 1800

It rained almost incessantly at Keswick till the late evening, when
it fell a deep calm, and even the leaves, and very topmost leaves,
of the poplars and aspens had holiday, and like an overworked
boy consumed it in sound sleep. . . . The clouds were scattered
by wind and rain in all shapes and heights, above the mountains,
on their sides, and low down to their bases—some masses in the
middle of the valley—when the wind and rain dropt down and
died, and for two hours all the clouds, white and fleecy all of
them, remained without motion, forming an appearance not very
unlike the moon as seen thro' a telescope. . . . Blessings on the
Mountains! to the eye and ear they are always faithful. I have
often thought of writing a set of play-bills for the Vale of Keswick—
for every day in the year—announcing each day the performance,
by his supreme Majesty's Servants, Clouds, Waters, Sun, Moon,
Stars, etc.

SAMUEL COLERIDGE, 26 July 1802

The *white rose* of Eddy-foam, where the stream ran into a scooped
or scolloped hollow of the Rock in its channel—this shape, an
exact white rose, was for ever overpowered by the Stream rushing down in upon it, and still obstinate in resurrection it spread
up into the Scollop, by fits and starts, *blossoming* in a moment
into full Flower.—Hung over the Bridge, & musing considering
how much of this Scene of endless variety in Identity was Nature's—
how much the living organ's!—What would it be if I had the eyes
of a fly!—what if the blunt eye of a Brobdignag!—

Black round Ink-spots from 5 to 18 in the decaying Leaf of the
Sycamore.

A circular glade in a forest of Birch Trees, and in the center
of the circle, a stone standing upright, twice a tall man's Height—
and by its side a stately Ash Tree umbrellaing it.—

A road on the breast of the mountain, all wooded save at the
very Top where the steep naked Crag lorded it—this road seen

only by a stream of white Cows, gleaming behind the Trees, in the Interspaces.

A Host of little winged Flies on the Snow mangled by the Hail Storm, near the Top of Helvellin.

SAMUEL COLERIDGE, misc. notes, 1803

As the comparatively small size of the lakes in the North of England is favourable to the production of variegated landscape, their *boundary-line* also is for the most part gracefully or boldly indented. That uniformity which prevails in the primitive frame of the lower grounds among all chains or clusters of mountains where large bodies of still water are bedded, is broken by the *secondary* agents of nature, ever at work to supply the deficiencies of the mould in which things were originally cast. Using the word *deficiencies*, I do not speak with reference to those stronger emotions which a region of mountains is peculiarly fitted to excite. The bases of those huge barriers may run for a long space in straight lines, and these parallel to each other; the opposite sides of a profound vale may ascend as exact counterparts, or in mutual reflection, like the billows of a troubled sea; and the impression be, from its very simplicity, more awful and sublime. Sublimity is the result of Nature's first great dealings with the superficies of the earth; but the general tendency of her subsequent operations is towards the production of beauty, by a multiplicity of symmetrical parts uniting in a consistent whole. This is every where exemplified along the margins of these lakes. Masses of rock, that have been precipitated from the heights into the area of waters, lie in some places like stranded ships; or have acquired the compact structure of jutting piers; or project in little peninsulas crested with native wood. . . . let us again recur to Nature. The process, by which she forms woods and forests, is as follows. Seeds are scattered indiscriminately by winds, brought by waters, and dropped by birds. They perish, or produce, according as the soil and situation upon which they fall are suited to them: and under the same dependence, the seedling or the sucker, if not cropped by animals, (which Nature is often careful to prevent by fencing it about with brambles or other prickly shrubs) thrives, and the tree grows, sometimes single, taking its own shape without constraint, but for the most part compelled to conform itself to some law imposed upon it by its neighbours. From low and sheltered places, vegetation travels upwards to the more exposed; and the young plants are protected, and to a certain degree fashioned, by those that have preceded them. The continuous mass of foliage which would be thus produced, is broken by rocks, or by glades or open places, where the browzing of animals has prevented the growth of wood. As vegetation ascends, the winds begin also to bear their part in moulding the forms of the trees; but, thus mutually protected, trees, though not of the hardiest kind, are enabled to climb high up the mountains. Gradually, however, by the quality of the ground, and by increasing exposure, a stop is put to their ascent; the hardy trees only are left: those also, by little and little, give way—and a wild and irregular boundary is established, graceful in its outline, and never contemplated without some feeling, more or less distinct, of the powers of Nature by which it is imposed.

Contrast the liberty that encourages, and the law that limits, this joint work of nature and time, with the disheartening necessities, restrictions, and disadvantages, under which the artificial planter must proceed, even he whom long observation and fine feeling have best qualified for his task. In the first place his trees, however well chosen and adapted to their several situations, must generally start all at the same time; and this necessity would of itself prevent that fine connection of parts, that sympathy and organization, if I may so express myself, which pervades the whole of a natural wood, and appears to the eye in its single trees, its masses of foliage, and their various colours, when they are held up to view on the side of a mountain; or when, spread over a valley, they are looked down upon from an eminence. It is therefore impossible, under any circumstances, for the artificial planter to rival the beauty of nature. But a moment's thought will show that, if ten thousand of this spiky tree, the larch, are stuck in at once upon the side of a hill, they can grow up into nothing but deformity; that, while they are suffered to stand, we shall look in vain for any of those appearances which are the chief sources of beauty in a natural wood.

WILLIAM WORDSWORTH, *Description of the Scenery of the Lakes*, 1810

———

I do not know that any one has ever explained satisfactorily the true source of our attachment to natural objects, or of that soothing emotion which the sight of the country hardly ever fails to infuse into the mind. Some persons have ascribed this feeling to the natural beauty of the objects themselves, others to the freedom from care, the silence and tranquillity which scenes of retirement afford—others to the healthy and innocent employments of a country life—others to the simplicity of country manners—and others to different causes; but none to the right one. All these causes may, I believe, have a share in producing this feeling; but there is another more general principle, which has been left untouched, and which I shall here explain, endeavouring to be as little sentimental as the subject will admit . . .

Were it not for the recollections habitually associated with them, natural objects could not interest the mind in the manner they do. No doubt, the sky is beautiful; the clouds sail majestically along its bosom; the sun is cheering; there is something exquisitely graceful in the manner in which a plant or tree puts forth its branches; the motion with which they bend and tremble in the evening breeze is soft and lovely; there is music in the babbling of a brook; the view from the top of a mountain is full of grandeur; nor can we behold the ocean with indifference . . .

It is not, however, the beautiful and magnificent alone that we admire in Nature; the most insignificant and rudest objects are often found connected with the strongest emotions; we become attached to the most common and familiar images as to the face of a friend whom we have long known, and from whom we have received many benefits. It is because natural objects have been associated with the sports of our childhood, with air and exercise, with our feelings in solitude, when the mind takes the strongest hold of things, and clings with the fondest interest to whatever strikes its attention; with change of place, the pursuit of new scenes, and thoughts of distant friends; it is because they have surrounded us in almost all situations, in joy and in sorrow, in pleasure and in pain; because they have been one chief source and nourishment of our feelings, and a part of our being, that we love them as we do ourselves.

There is, generally speaking, the same foundation for our love of Nature as for all our habitual attachments, namely, association of ideas. But this is not all. That which distinguishes this attachment from others is the transferable nature of our feelings with respect to physical objects; the associations connected with any one object extending to the whole class. My having been attached to any particular person does not make me feel the same attachment to the next person I may chance to meet; but, if I have once associated strong feelings of delight with the objects of natural scenery, the tie becomes indissoluble, and I shall ever after feel the same attachment to other objects of the same sort. I remember when I was abroad, the trees, and grass, and wet leaves, rustling in the walks of the Thuilleries, seemed to be as much English, to be as much the same trees and grass, that I had always been used to, as the sun shining over my head was the same sun which I saw in England; the faces only were foreign to me. Whence comes this difference? It arises from our always imperceptibly connecting the idea of the individual with man, and only the idea of the class with natural objects. In the one case, the external appearance or physical structure is the least thing to be attended to; in the other, it is every thing. The springs that move the human form, and make it friendly or adverse to me, lie hid within it. There is an infinity of motives, passions, and ideas, contained in that narrow compass, of which I know nothing, and in which I have no share. Each individual is a world to himself, governed by a thousand contradictory and wayward impulses. I can, therefore, make no inference from one individual to another; nor can my habitual sentiments, with respect to any individual, extend beyond himself to others. But it is otherwise with respect to Nature. There is neither hypocrisy, caprice, nor mental reservation in her favours. Our intercourse with her is not liable to accident or change, interruption or disappointment. She smiles on us still the same. Thus, to give an obvious instance, if I have once enjoyed the cool shade of a tree, and been lulled into a deep repose by the sound of a brook running at its feet, I am sure that wherever I can find a tree and a brook, I can enjoy the same pleasure again. Hence, when I imagine these objects, I can easily form a mystic personification of the friend-ly power that inhabits them, Dryad or Naiad, offering its cool fountain or its tempting shade. Hence the origin of the Grecian mythology. All objects of the same kind being the same, not only in their appearance, but in their practical uses, we habitually con-found them together under the same general idea; and, what-ever fondness we may have conceived for one, is immediately placed to the common account. The most opposite kinds and remote trains of feeling gradually go to enrich the same sentiment; and in our love of Nature, there is all the force of individual attachment, combined with the most airy abstraction. It is this circumstance which gives that refinement, expansion, and wild interest to feelings of this sort, when strongly excited, which every one must have experienced who is a true lover of Nature. The sight of the setting sun does not affect me so much from the beauty of the object itself, from the glory kindled through the glowing skies, the rich broken columns of light, or the dying streaks of day, as that it indistinctly recalls to me numberless thoughts and feelings with which, through many a year and season, I have watched his bright descent in the warm summer evenings, or beheld him struggling to cast a 'farewel sweet' through the thick clouds of winter. I love to see the trees first covered with leaves in the spring, the primroses peeping out from some sheltered bank, and the innocent lambs running races on the soft green turf; because, at that birth-time of Nature, I have always felt sweet hopes and happy wishes—which have not been fulfilled! The dry reeds rustling on the side of a stream,—the woods swept by the loud blast,—the dark massy foliage of autumn,—the grey trunks and naked branches of the trees in winter,—the sequestered copse and wide extended heath,—the warm sunny showers, and December snows,—have all charms for me; there is no object, however trifling or rude, that has not, in some mood or other, found the way to my heart . . . Thus Nature is a kind of universal home, and every object it presents to us an old acquaintance with unaltered looks. For there is that consent and mutual harmony among all her works, one undivided spirit pervading them throughout, that, if we have once knit ourselves in hearty fellowship to any of them, they will never afterwards appear as strangers to us, but, which ever way we turn, we shall find a secret power to have gone out before us, moulding them into such shapes as fancy loves, informing them with life and sympathy, bidding them put on their festive looks and gayest attire at our approach, and to pour all their sweets and choicest treasures at our feet. For him, then, who has well acquainted himself with Nature's works, she wears always one face, and speaks the same well-known language, striking on the heart, amidst unquiet thoughts and the tumult of the world, like the music of one's native tongue heard in some far-off country.

WILLIAM HAZLITT, 'On the Love of the Country', *The Examiner*, November 1814

THE BLACKTHORN

The tree, or rather the bush, on the subject of which I am now about to hope for the reader's attention, is pretty well known to most English people, who will generally, perhaps, look upon it as something of little importance; but which is of real importance as to the two great purposes to which it is applied; namely, the making of excellent *hedges*, and the making of excellent *Port wine*: in which last of its functions I shall consider it first.

Everyone knows that this is a Thorn of the Plum kind; that it bears very small black plums, which are called Sloes, which have served love-song poets, in all ages, with a simile whereby to describe the eyes of their beauties, as the snow has constantly served them with the means of attempting to do something like justice to the colour of their skins and the purity of their minds, and as the rose has served to assist them in a description of the colour of their cheeks.

These beauty-describing sloes, have a little plum-like pulp, which covers a little roundish stone, pretty nearly as hard as iron, with a small kernel in the inside of it. This pulp, which I have eaten many times when I was a boy until my tongue clove to the roof of my mouth and my lips were pretty nearly glued together, is astringent beyond the powers of alum. The juice expressed from this pulp is of a greenish black, and mixed with water, in which a due proportion of logwood has been steeped, receiving, in addition, a sufficient proportion of cheap French brandy, makes the finest Port wine in the world, makes the whiskered bucks, while they are picking their teeth after dinner, smack their lips observing that the wine is beautifully rough and that they like 'a *dry wine* that has a good *body*'.

It is not, however, as a fruit-tree that I am here about to speak seriously to sensible people; it is of a *bush* excellent for the making of *hedges*, and not less excellent for the making of walking sticks and swingles of flails. The Black Thorn blows very early in the spring. It is a Plum and it blows at the same time, or a very little earlier, than the Plums. It is a remarkable fact that there is always, that is every year of our lives, a spell of cold and angry weather just at the time that this hardy little tree is in bloom. The country people call it the *Black Thorn winter* and thus it has been called, I dare say by all the inhabitants of this island, from generation to generation, for a thousand years.

This thorn is as hardy as the White Thorn; its thorns are sharper and longer; it grows as fast; its wood is a great deal harder and more tough; it throws out a great deal more in side-shoots; and it is, in every respect, better than the Hawthorn for the making of a Hedge. If I be asked how it has happened, then, that the Hawthorn is constantly used for this purpose, and the Black Thorn never, or scarcely ever, I answer that the reason is very clear; namely that a sack of the seed of a Hawthorn may, almost anywhere, be got for a shilling or half a crown at the most; and that, to get a number of Black Thorn sloes, equal in number to the Hawthorn berries contained in a sack, would, in almost any part of the kingdom, cost five, ten, nay twenty pounds.

The sloe is very large compared with the size of the Hawthorn berry; you must get six sacks perhaps of the sloes to have a number equal to the berries contained in one sack; and six sacks of sloes, except in very woody countries, would not be found perhaps in the half of a whole county. The tree, like other plums, is liable to blight. It seldom bears any considerable crop, and very frequently bears no fruit at all. It grows no where except in hedge-rows and coppices; in the former it is too much exposed to bear much fruit; and in the latter, it is too much in the shade to bear any fruit at all. Hence it is, that, though all of us who have been born and bred in the country know that the Blackthorn is by far the best of the two, we never heard of such a thing as the planting of a Black Thorn Hedge.

As *bushes*, for the making of hedges, the Black Thorn is always carefully laid by when hedge-rows and coppices are cut. These bushes will lay longer in a dead hedge without perishing than any other sort of stuff of which dead hedges are made. This Thorn will thrive, and that vigorously too, in the very poorest of land. It sends up straighter shoots from the stem than the White Thorn does, and these shoots send out, from their very bottom, numerous and vigorous side-shoots, all armed with sharp thorns. The knots produced by these side-shoots are so thickly set, that, when the shoot is cut, whether it be little or big, it makes the most beautiful of all walking or riding sticks. The bark, which is precisely of the colour of the Horse Chestnut fruit and as smooth and as bright, needs no polish; and, ornamented by the numerous knots, the stick is the very prettiest that can be conceived. Little do the bucks, when they are drinking Port wine (good old rough Port) imagine that, by possibility, the beautiful stick with which they are tapping the sole of their boot, while admiring their legs, never does their philosophy carry them so far as to lead them to reflect that, by possibility, for the 'fine old Port,' which has caused them so much pleasure, they are indebted to the very stick with which they are caressing their admired Wellington boots.

WILLIAM COBBETT, *The Woodlands*, 1825. Traveller, farming writer, and radical polemicist, Cobbett is not normally included amongst the Romantics. But his physical relish in the natural world gives him more links with them than might be imagined from the bluff surfaces of his writing. Hazlitt commended his prose as 'plain, broad, downright English'

It has been a commonly believd notion among such naturalists that trusts to books & repeats old error that ants hurd up & feed on the curnels of grain such as wheat & barley but every common observer knows this to be a falshood I have noticd them minutly & often & never saw one with such food in its mouth they feed on flyes & catepillars which I have often seen them lugging home with & for which they climb trees & the stems of flowers—when they first appear in the spring they may be seen carring out ants in their mouths of a smaller size which they will continue to do a long time transporting them away from home perhaps to form new colonys perhaps—they always make one track from their nest & keep it & will go for furlongs away from their homes fetching bits of bents & others lugging away with flyes or green maggots which they pick off of flowers & leaves some when over loaded are joind by others till they get a sufficient quantity to master it home I have often minded that two while passing each other woud pause like old friends long seperated & as if they suddenly recollectd each other they went & put their heads together as if they shook hands or saluted each other when a shower comes on an unusual bustle ensues round their nest some set out & suddenly turn round again without fetching any thing home others will hasten to help those that are loaded & when the rain begins to fall the others will leave their loads & make the best of their ways as fast as they can their general employment is the gathering of bents &c to cover their habitation which they generaly make round an old root which they eat into holes like an honey comb these holes lead & comunicate into each other for a long way in the ground in winter they lye dormant but quickly revive if exposd to the sun there is nothing

to be seen of food in their habitations then—I have observd strag-
glers that crawl about the grass \ seemlingly without a purpose / &
if they accidentialy fall into the track of those at labor that quick-
en their pace & suddenly retreat I have fancied these to be the
idle & discontented sort of radicals to the government & larger
of the rest

the smaller ants calld pismires seem to be under a different
sort of government at least there is not that regularity observed
among them in their labour as there is among the large ones they
do not keep one track as the others do but creep about the grass
were they please they are uncommonly fond of bread (which the
larger ants will not touch) & when the shepherd litters his crumbles from his dinner bag on their hill as he often will to observe
them it instantly creates a great bustle among the little colony &
they hasten away with it as fast as they can till every morsel is
cleand up when they pause about as if looking for more—it is
commonly believed by carless observers that every hillock on
greens & commons has been first rooted up & afterwards occupied by these little tenants but on the contrery most of the hills
they occupy are formed by themselves which they increase every
year by bringing up a portion of mould on the survace finely
powderd in which they lay their eggs to recieve the warmth of
the sun & the shepherd by observing their wisdom in this labour
judges corectly of the changes of the weather in fact he finds it
an infallable almanack when fine weather sets in their eggs are
brought nearly to the top of this new addition to their hill & as
soon as ever a change is about to take place nay at the approach
of a shower they are observd carrying them deeper down to safer
situations & if much wet is coming they entirely dissapear with
them into the bosom of the castles were no rain can reach them
for they generally use a composition of clay in making the hills
that forces off the wet & keeps it from penetrating into their cells
if \ the crown of / one of these hills but taken off with a spade it
will appear pierced with holes like a honey comb—these little
things are armd with stings that blister & torture the skin with a
pain worse then the keen nettle there is a smaller sort still of a
deep black color that like the large ones have no sting I once
when sitting at my dinner hour in the fields seeing a colony of
the red pismires near one of these black ones tryd the experiment to see wether they woud assosiate with each other & as
soon as I put a black one among them they began to fight when

the latter after wounding his antagonist (seeming\ always / to be of inferior strength) curled up & dyd at his feet I then put a red one to the colony of blacks which they instantly seized & tho he generally contrived to escape he appeard to be terribly wounded & no doubt was a cripple for life—these little creatures will raise a large tower of earth as thick as a mans arm in the form of a sugar loaf to a foot & a foot & a half high in the grain & long grass—for in such places they cannot meet the sun on the ground so they raise these towers in the top of which they lay their eggs & as the grass or grain keeps growing they keep raising their towers till I have met with them as tall as ones knees

> JOHN CLARE, 'On Ants', c.1824/5: from *The Natural History Prose Writings of John Clare*, ed. Margaret Grainger, 1983. John Clare began assembling natural history prose notes in the mid-1820s at the suggestion of his publisher, John Taylor, who hoped that he might follow in Gilbert White's footsteps with *A Natural History of Helpstone* (Clare's home village near Peterborough). The book was never completed

10 Sept. 1824

The swallows are flocking together in the skies ready for departing and a crowd has dropt to rest on the walnut tree where they twitter as if they were telling their young stories of their long journey to cheer and check fears.

*

Saturday 25 Dec. 1824

Christmass\ day / gatherd a hand ful of daiseys in full bloom—saw a wood bine & dog rose in the woods putting out in full leaf & a primrose root full of ripe flowers—what a day this usd to be when a boy how eager I usd to attend the church to see it stuck with evergreens (emblems of Eternity) & the cottage windows & the picture ballads on the wall all stuck with Ivy holly Box & yew—such feelings are past—& 'all this world is proud of'

*

Sunday 26 Dec. 1824

Found at the bottom of a dyke made in the roman bank some pootys of varied colors & the large garden ones of a russet color with a great many others of the meadow sort which we calld 'badgers' when I was a school boy found no were now but in wet places—there is a great many too of a water species now extinct—the dyke is 4 foot deep & the soil is full of these shells— have they not lay here ever since the romans made the bank & does the water sorts not imply that the fields was all fen & under water or wet & uncultivated at that time I think it does—I never walk on this bank but the legions of the roman army pass bye my fancys with their mysterys of nearly 2000 years hanging like a mist around them what changes hath past since then—were I found these shells it was heath land above 'swordy well'

> JOHN CLARE, *Journal* entries 1824: from *The Natural History Prose Writings of John Clare*, ed. Margaret Grainger, 1983

[*5 July 1852*]

Some birds are poets and sing all summer. They are the true singers. Any man can write verses during the love season. I am reminded of this while we rest in the shade on the Major Heywood road and listen to a wood thrush, now just before sunset. We are most interested in those birds who sing for the love of the music and not of their mates; who meditate their strains, and *amuse* themselves with singing; the birds, the strains, of deeper senti- ment; not bobolinks, that lose their plumage, their bright colors, and their song so early.

The robin, the red-eye, the veery, the wood thrush, etc., etc.

The wood thrush's is no opera music; it is not so much the composition as the strain, the tone,—cool bars of melody from the atmosphere of everlasting morning or evening. It is the qual- ity of the song, not the sequence. In the peawai's note there is some sultriness, but in the thrush's, though heard at noon, there is the liquid coolness of things that are just drawn from the bot- tom of springs. The thrush alone declares the immortal wealth and vigor that is in the forest. Here is a bird in whose strain the story is told, though Nature waited for the science of æsthetics to discover it to man. Whenever a man hears it, he is young, and Nature is in her spring. Wherever he hears it, it is a new world and a free country, and the gates of heaven are not shut against him. Most other birds sing from the level of my ordinary cheer- ful hours—a carol; but this bird never fails to speak to me out of an ether purer than that I breathe, of immortal beauty and vigor. He deepens the significance of all things seen in the light of his strain. He sings to make men take higher and truer views of things. He sings to amend their institutions; to relieve the slave on the plantation and the prisoner in his dungeon, the slave in the house of luxury and the prisoner of his own low thoughts.

*

Dec. 11. P. M.—To Holden Swamp, Conantum. For the first time I wear gloves, but I have not walked *early* this season.

I see no birds, but hear, methinks, one or two tree sparrows. No snow; scarcely any ice to be detected. It is only an aggravated November. I thread the tangle of the spruce swamp, admiring the leafets of the swamp pyrus which had put forth again, now frostbitten, the great yellow buds of the swamp-pink, the round red buds of the high blueberry, and the fine sharp red ones of the panicled andromeda. Slowly I worm my way amid the snarl, the thicket of black alders and blueberry, etc.; see the forms, apparently, of rabbits at the foot of maples, and catbirds' nests now exposed in the leafless thicket.

Standing there, though in this *bare* November landscape, I am reminded of the incredible phenomenon of small birds in winter. That ere long, amid the cold powdery snow, as it were a fruit of the season, will come twittering a flock of delicate crimson-tinged birds, lesser redpolls, to sport and feed on the seeds and buds now just ripe for them on the sunny side of a wood, shaking down the powdery snow there in their cheerful social feeding, as if it were high midsummer to them. These crimson aerial creatures have wings which would bear them quickly to the regions of summer, but here is all the summer they want. What a rich contrast! tropical colors, crimson breasts, on cold white snow! Such etherealness, such delicacy in their forms, such ripeness in their colors, in this stern and barren season! It is as surprising as if you were to find a brilliant crimson flower which flourished amid snows . . .

When some rare northern bird like the pine grosbeak is seen thus far south in the winter, he does not suggest poverty, but dazzles us with his beauty. There is in them a warmth akin to the warmth that melts the icicle. Think of these brilliant, warm-colored, and richly warbling birds, birds of paradise, dainty-footed, downy-clad, in the midst of a New England, a Canadian winter.

The woods and fields, now somewhat solitary, being deserted by their more tender summer residents, are now frequented by these rich but delicately tinted and hardy northern immigrants of the air. Here is no imperfection to be suggested. The winter, with its snow and ice, is not an evil to be corrected. It is as it was designed and made to be, for the artist has had leisure to add beauty to use. My acquaintances, angels from the north. I had a vision thus prospectively of these birds as I stood in the swamps. I saw this familiar—too *familiar*—fact at a different angle, and I was charmed and haunted by it. But I could only attain to be thrilled and enchanted, as by the sound of a strain of music dying away. I had seen into paradisaic regions, with their air and sky, and I was no longer wholly or merely a denizen of this vulgar earth. Yet had I hardly a foothold there. I was only sure that I was charmed, and no mistake. It is only necessary to behold thus the least fact or phenomenon, however familiar, from a point a hair's breadth aside from our habitual path or routine, to be overcome, enchanted by its beauty and significance. Only what we have touched and worn is trivial,—our scurf, repetition, tradition, conformity. To perceive freshly, with fresh senses, is to be inspired. Great winter itself looked like a precious gem, reflecting rainbow colors from one angle.

My body is all sentient. As I go here or there, I am tickled by this or that I come in contact with, as if I touched the wires of a battery. I can generally recall—have fresh in my mind—several scratches last received. These I continually recall to mind, reimpress, and harp upon. The age of miracles is each moment thus returned. Now it is wild apples, now river reflections, now a flock of lesser redpolls. In winter, too, resides immortal youth and perennial summer. Its head is not silvered; its cheek is not blanched but has a ruby tinge to it.

If any part of nature excites our pity, it is for ourselves we grieve, for there is eternal health and beauty. We get only transient and partial glimpses of the beauty of the world. Standing at the right angle, we are dazzled by the colors of the rainbow in colorless ice. From the right point of view, every storm and every drop in it is a rainbow. Beauty and music are not mere traits and exceptions. They are the rule and character. It is the exception that we see and hear. Then I try to discover what it was in the vision that charmed and translated me. What if we could daguerreotype our thoughts and feelings! for I am surprised and enchanted often by some quality which I cannot detect. I have seen an attribute of another world and condition of things. It is a wonderful fact that I should be affected, and thus deeply and powerfully, more than by aught else in all my experience,— that this fruit should be borne in me, sprung from a seed finer than the spores of fungi, floated from other atmospheres! finer than the dust caught in the sails of vessels a thousand miles from land! Here the invisible seeds settle, and spring, and bear flowers and fruits of immortal beauty.

HENRY THOREAU, *Journals*, 5 July 1852 and 11 Dec. 1855

The other day, as I was calling on the ornithologist whose collection of birds is, I suppose, altogether unrivalled in Europe,— (at once a monument of unwearied love of science, and an example, in its treatment, of the most delicate and patient art)—Mr. Gould—he showed me the nest of a common English bird; a nest which, notwithstanding his knowledge of the dexterous building of birds in all the world, was not without interest even to him, and was altogether amazing and delightful to me. It was a bullfinch's nest, which had been set in the fork of a sapling tree, where it needed an extended foundation. And the bird had built this first story of her nest with withered stalks of clematis blossom; and with nothing else. These twigs it had interwoven lightly, leaving the branched heads all at the outside, producing an intricate Gothic boss of extreme grace and quaintness, apparently arranged both with triumphant pleasure in the art of basket-making, and with definite purpose of obtaining ornamental form.

I fear there is no occasion to tell you that the bird had no purpose of the kind. I say that I *fear* this, because I would much rather have to undeceive you in attributing too much intellect to the lower animals, than too little. But I suppose the only error which, in the present condition of natural history, you are likely to fall into, is that of supposing that a bullfinch is merely a mechanical arrangement of nervous fibre, covered with feathers by a chronic cutaneous eruption; and impelled by a galvanic stimulus to the collection of clematis.

You would be in much greater, as well as in a more shameful, error, in supposing this, than if you attributed to the bullfinch the most deliberate rivalship with Mr. Street's prettiest Gothic designs. The bird has exactly the degree of emotion, the extent of science, and the command of art, which are necessary for its happiness; it had felt the clematis twigs to be lighter and tougher than any others within its reach, and probably found the forked branches of them convenient for reticulation. It had naturally placed these outside, because it wanted a smooth surface for the bottom of its nest; and the beauty of the result was much more dependent on the blossoms than the bird.

Nevertheless, I am sure that if you had seen the nest,—much more, if you had stood beside the architect at work upon it,— you would have greatly desired to express your admiration to her; and that if Wordsworth, or any other simple and kindly person, could even wish, for a little flower's sake,

> 'That to this mountain daisy's self were known
> The beauty of its star-shaped shadow, thrown
> On the smooth surface of this naked stone,'

much more you would have yearned to inform the bright little nest-builder of your sympathy; and to explain to her, on art principles, what a pretty thing she was making.

Does it never occur to you, then, that to some of the best and wisest artists among ourselves, it may not be always possible to explain what pretty things they are making; and that, perhaps, the very perfection of their art is in their knowing so little about it?

Whether it has occurred to you or not, I assure you that it is so. The greatest artists, indeed, will condescend, occasionally, to be scientific;—will labour, somewhat systematically, about what they are doing, as vulgar persons do; and are privileged, also, to enjoy what they have made more than birds do; yet seldom, observe you, as being beautiful, but very much in the sort of feeling which we may fancy the bullfinch had also,—that the thing, whether pretty or ugly, could not have been better done; that they could not have made it otherwise, and are thankful it is no worse. And, assuredly, they have nothing like the delight in their own work which it gives to other people.

JOHN RUSKIN, 'The Relation of Wise Art to Wise Science', *The Eagle's Nest*, 1887

The morning suddenly became glorious and we saw what had happened in the night. All night long millions of gossamer spiders had been spinning and the whole country was covered. . . . The gossamer webs gleamed and twinkled into crimson and gold and green, like the most exquisite shot-silk dress in the finest texture of gauzy silver wire. I never saw anything like it or anything so exquisite as 'the Virgin's webs' glowed with changing opal lights and glanced with all the colours of the rainbow. At 4 o'clock Miss Meredith Brown and her beautiful sister Etty came over to afternoon tea with us and a game of croquet.

> Francis Kilvert, *Diary*, 6 Sept. 1875, ed. William Plomer, 1944

This day and May 11 the bluebells in the little wood between the College and the highroad and in one of the Hurst Green cloughs. In the little wood / opposite the light / they stood in black-ish spreads or sheddings like the spots on a snake. The heads are then like thongs and solemn in grain and grape-colour. But in the clough / through the light / they came in falls of sky-colour washing the brows and slacks of the ground with vein-blue, thickening at the double, vertical themselves and the young grass and brake fern combed vertical, but the brake struck the upright of all this with light winged transoms. It was a lovely sight.—The bluebells in your hand baffle you with their inscape, made to every sense: if you draw your fingers through them they are lodged and struggle / with a shock of wet heads; the long stalks rub and click and flatten to a fan on one another like your fingers themselves would when you passed the palms hard across one another, making a brittle rub and jostle like the noise of a hurdle strained by leaning against; then there is the faint honey smell and in the mouth the sweet gum when you bite them. But this is easy, it is the eye they baffle. They give one a fancy of pan-pipes and of some wind instrument with stops—a trombone perhaps. The overhung necks—for growing they are little more than a staff with a simple crook but in water, where they stiffen, they take stronger turns, in the head like sheephooks or, when more waved throughout, like the waves riding through a whip that is being smacked—what with these overhung necks and what with the crisped ruffled bells dropping mostly on one side and the gloss these have at their footstalks they have an air of the knights at chess. Then the knot or 'knoop' of buds some shut, some just gaping, which makes the pencil of the whole spike, should be noticed: the inscape of the flower most finely carried out in the siding of the axes, each striking a greater and greater slant, is finished in these clustered buds, which for the most part are not straightened but rise to the end like a tongue and this and their tapering and a little flattening they have make them look like the heads of snakes.

GERARD MANLEY HOPKINS, *Journals*, 9 May 1871

It is sweet on awaking in the early morn to listen to the small bird singing on the tree. No sound of voice or flute is like the bird's song; there is something in it distinct and separate from all other notes. The throat of woman gives forth a more perfect music, and the organ is the glory of man's soul. The bird upon the tree utters the meaning of the wind—a voice of the grass and wild flower, words of the green leaf; they speak through that slender tone. Sweetness of dew and rifts of sunshine, the dark hawthorn touched with breadths of open bud, the odour of the air, the colour of the daffodil—all that is delicious and beloved of springtime are expressed in his song. Genius is nature, and his lay, like the sap in the bough from which he sings, rises without thought. Nor is it necessary that it should be a song; a few short notes in the sharp spring morning are sufficient to stir the heart. But yesterday the least of them all came to a bough by my window, and in his call I heard the sweet-briar wind rushing over the young grass. Refulgent fall the golden rays of the sun; a minute only, the clouds cover him and the hedge is dark. The bloom of the gorse is shut like a book; but it is there—a few hours of warmth and the covers will fall open. The meadow is bare, but in a little while the heart-shaped celandine leaves will come in their accustomed place. On the pollard willows the long wands are yellow-ruddy in the passing gleam of sunshine, the first colour of spring appears in their bark. The delicious wind rushes among them and they bow and rise; it touches the top of the dark pine that looks in the sun the same now as in summer; it lifts and swings the arching trail of bramble; it dries and crumbles the earth in its fingers; the hedge-sparrow's feathers are fluttered as he sings on the bush.

I wonder to myself how they can all get on without me—how they manage, bird and flower, without me to keep the calendar for them. For I noted it so carefully and lovingly, day by day,

the seed-leaves on the mound in the sheltered places that come
so early, the pushing up of the young grass, the succulent dandelion, the coltsfoot on the heavy, thick clods, the trodden chick-
weed despised at the foot of the gate-post, so common and small,
and yet so dear to me. Every blade of grass was mine, as though
I had planted it separately. They were all my pets, as the roses
the lover of his garden tends so faithfully. All the grasses of the
meadow were my pets, I loved them all; and perhaps that was
why I never had a 'pet', never cultivated a flower, never kept a
caged bird, or any creature. Why keep pets when every wild free
hawk that passed overhead in the air was mine? I joyed in his
swift, careless flight, in the throw of his pinions, in his rush over
the elms and miles of woodland; it was happiness to see his
unchecked life. What more beautiful than the sweep and curve
of his going through the azure sky? These were my pets, and all
the grass. Under the wind it seemed to dry and become grey,
and the starlings running to and fro on the surface that did not
sink now stood high above it and were larger. The dust that drifted
along blessed it and it grew. Day by day a change; always a note
to make. The moss drying on the tree trunks, dog's-mercury stirring under the ash-poles, bird's-claw buds of beech lengthening;
books upon books to be filled with these things. I cannot think
how they manage without me.

To-day through the window-pane I see a lark high up against
the grey cloud, and hear his song. I cannot walk about and arrange
with the buds and gorse-bloom; how does he know it is the time
for him to sing? Without my book and pencil and observing eye,
how does he understand that the hour has come? To sing high
in the air, to chase his mate over the low stone wall of the
ploughed field, to battle with his high-crested rival, to balance
himself on his trembling wings outspread a few yards above the
earth, and utter that sweet little loving kiss, as it were, of song—
oh, happy, happy days! So beautiful to watch as if he were my
own, and I felt it all! It is years since I went out amongst them
in the old fields, and saw them in the green corn, they must be
dead, dear little things, by now. Without me to tell him, how
does this lark to-day that I hear through the window know it is
his hour?

The green hawthorn buds prophesy on the hedge; the reed
pushes up in the moist earth like a spear thrust through a shield;
the eggs of the starling are laid in the knot-hole of the pollard

elm—common eggs, but within each a speck that is not to be found in the cut diamond of two hundred carats—the dot of protoplasm, the atom of life. There was one row of pollards where they always began laying first. With a big stick in his beak the rook is blown aside like a loose feather in the wind; he knows his building-time from the fathers of his house—hereditary knowledge handed down in settled course: but the stray things of the hedge, how do they know? The great blackbird has planted his nest by the ash-stole, open to every one's view, without a bough to conceal it and not a leaf on the ash—nothing but the moss on the lower end of the branches. He does not seek cunningly for concealment. I think of the drift of time, and I see the apple bloom coming and the blue veronica in the grass. A thousand thousand buds and leaves and flowers and blades of grass, things to note day by day, increasing so rapidly that no pencil can put them down and no book hold them, not even to number them— and how to write the thoughts they give? All these without me— how can they manage without me?

For they were so much to me, I had come to feel that I was as much in return to them. The old, old error: I love the earth, therefore the earth loves me—I am her child—I am Man, the favoured of all creatures. I am the centre, and all for me was made.

In time past, strong of foot, I walked gaily up the noble hill that leads to Beachy Head from Eastbourne, joying greatly in the sun and the wind. Every step crumbled up numbers of minute grey shells, empty and dry, that crunched under foot like hoarfrost or fragile beads. They were very pretty; it was a shame to crush them—such vases as no king's pottery could make. They lay by millions in the depths of the sward, and I thought as I broke them unwillingly that each of these had once been a house of life. A living creature dwelt in each and felt the joy of existence, and was to itself all in all—as if the great sun over the hill shone for it, and the width of the earth under was for it, and the grass and plants put on purpose for it. They were dead, the whole race of them, and these their skeletons were as dust under my feet. Nature sets no value upon life neither of minute hill-snail nor of human being.

I thought myself so much to the earliest leaf and the first meadow orchis—so important that I should note the first zee-zee of the titlark—that I should pronounce it summer, because now the oaks were green; I must not miss a day nor an hour in the fields lest something should escape me. How beautiful the droop of the great brome-grass by the wood! But to-day I have to listen to the lark's song—not out of doors with him, but through the window-pane, and the bullfinch carries the rootlet fibre to his nest with-out me. They manage without me very well; they know their times and seasons—not only the civilized rooks, with their libraries of knowledge in their old nests of reference, but the stray things of the hedge and the chiffchaff from over sea in the ash wood. They go on without me. Orchis flower and cowslip—I cannot number them all—I hear, as it were, the patter of their feet—flower and bud and the beautiful clouds that go over, with the sweet rush of rain and burst of sun glory among the leafy trees. They go on, and I am no more than the least of the empty shells that strewed the sward of the hill. Nature sets no value upon life, neither of mine nor of the larks that sang years ago. The earth is all in all to me, but I am nothing to the earth: it is bitter to know this before you are dead. These delicious violets are sweet for themselves; they were not shaped and coloured and gifted with that exquisite proportion and adjustment of odour and hue for me. High up against the grey cloud I hear the lark through the window singing, and each note falls into my heart like a knife.

RICHARD JEFFERIES, 'Hours of Spring', *Longman's Magazine*, May 1886: first collected in *Field and Hedgerow*, 1889

It was a beech, standing somewhat isolated, and still leafless in quite early spring. Suddenly I was aware of its skyward-reaching arms and upturned fingertips, as if some vivid life (or electrici-ty) was streaming through them into the spaces of Heaven, and of its roots plunged in the earth and drawing the same energies from below. The day was quite still and there was no movement in the branches, but in that moment the tree was no longer a separate or separable organism, but a vast being ramifying far into space, sharing and uniting the life of earth and sky, and full of a most amazing activity.

EDWARD CARPENTER, 'Pagan and Christian Creeds', 1905

Wonders of Creation

FROM the late eighteenth century much European natural history was driven by a sense of mission. There was ecological plunder abroad (thinly justified as a process of civilization) and awe at the Wonders of Creation at home. By the Victorian era natural history had become a public obsession. The ceaseless parade of natural marvels (like the Victoria Regia water-lily, big enough to need a conservatory to itself) brought back to Britain from the outposts of empire exactly suited the prevailing mood of expansiveness and optimism. So, in another way, did the beauty and pattern that could be glimpsed by God-fearing observers in the simplest native fern or shell. Both seemed like a divine blessing on the nation. It is no surprise to find that contemporary books such as The Sagacity and Morality of Plants *sold in hundreds of thousands.*

But there was real enthusiasm and curiosity behind the piety and self-congratulation, especially at moments of culture-shock. Naturalists found the bizarre natural life of the southern hemisphere especially challenging. It helped inspire von Humboldt's vision of the whole earth as an organism (anticipating James Lovelock's Gaia hypothesis, which sees the physical earth and the biosphere linked as a self-regulating system), and Darwin's theory of evolution. Some travellers, struggling to make sense of the plants and creatures of Australia, fancied they had discovered a type of Eden, where Noble Savages really existed; or an 'antipodal inversion' of the natural order as it was experienced in the north. But many coped with the disorientation of the bush by simply likening it to a 'gentleman's park' in England. However far-fetched and fabulous the wonders were, for the Victorians there was no touchstone like home.

This tract of country has afforded more riches for the naturalist than perhaps any other part of the globe. When the Europeans first settled there, the whole might have been compared to a great park, furnished with a wonderful variety of animals . . . but since the country has been inhabited by Europeans, most of these have been destroyed or driven away . . .

[20 November 1773] At night we got clear of the mountains, but entered a rugged country, which the new inhabitants name Canaan's Land; though it might rather be called the Land of Sorrow; for no land could exhibit a more wasteful prospect; the plains consisting of nothing but rotten rock, intermixed with a little red loam in the interstices, which supported a variety of scrubby bushes, in their nature evergreen, but, by the scorching heat of the sun, stripped almost of all their leaves. Yet notwithstanding the disagreeable aspect of this tract, we enriched our collection by a variety of succulent plants, which we had never seen before, and which appeared to us like a new creation . . .

[26 October 1774] The steril appearance of this country exceeds all imagination: wherever one casts his eyes he sees nothing but naked hills, without a blade of grass, only small succulent plants. The soil is a red binding loam, intermixed with a kind of rotten *schistus* or slate. Next morning we traversed the adjacent hills, and were surprized to find all the plants entirely new to us. They were the greatest part of the succulent kind; viz. *Mesembryanthemum*, *Euphorbia*, and *Stapelia*, of which we found many new species. The peasant told us, that in winter the hills were painted with all kind of colours; and said, it grieved him often, that no person of knowledge in botany had ever had an opportunity of seeing his country in the flowery season. We expressed great surprize at seeing such large flocks of sheep as he was possessed of subsist in such a desart; on which he observed, that their sheep never ate any grass, only succulent plants, and all sorts of shrubs; many of which were aromatic, and gave their flesh an excellent flavour. Next day I passed through a large flock of sheep, where I saw them devouring the juicy leaves of *Mesembryanthemum*, *Stapelia*, *Cotyledon*, and even the green seed vessels of *euphorbia*; by eating such plants they require little water, especially in winter.

[31 October] All next day we travelled over this thirsty land, where we suffered from the heat of the Sun and want of water; but our sufferings were still aggravated when we thought on our poor animals, who often lay down in the yoke during the heat of the day. This desart is extensive; being bounded on the N. and N.E. by a chain of flat mountains, called Bockland's Bergen

(Bockland's Mountains) and on the W. and N.W. by the Atlantic Ocean. It is uninhabitable in summer; but in winter, or during the rainy season, the Bockland people come down with their herds, which by feeding upon succulent shrubs, that are very salt, in a short time grow remarkably fat. There still remains a great treasure of new plants in this country, especially of the succulent kind, which cannot be preserved but by having good figures and descriptions of them made on the spot; which might be easily accomplished in the rainy season, when there is plenty of fresh water every where. But at this season of the year, we were obliged to make the greatest expedition to save the lives of our cattle, only collecting what we found growing along the road side, which amounted to above 100 plants, never before described.

> FRANCIS MASSON, *Philosophical Transactions of the Royal Society*, 1776. Francis Masson was a Scots under-gardener employed at Kew Gardens. His enthusiasm and botanical skills soon attracted the attention of Kew's unofficial director, Joseph Banks, and in 1772 he was despatched to the Cape of Good Hope to survey the local flora and collect plants. He proved to have a natural talent for plant illustration, and after twenty years in the field in southern Africa, produced a hauntingly beautiful collection of paintings of the region's succulent plants, entitled *Stapeliae Novae* (1797). His unassuming journal of his first expedition contains many sympathetic portraits of the local landscapes and wildlife, and how they were suffering at the hands of European colonists

Naturalists' reactions to first encounters with Australian plants and creatures.

When a botanist first enters on the investigation of so remote a country as New Holland, he finds himself as it were in a new world. He can scarcely meet with any certain fixed points from which to draw his analogies; and even those that appear most promising are frequently in danger of misleading him. Whole tribes of plants, which at first sight seem familiar to his acquaintance, as occupying links in Nature's chain . . . prove, on a nearer examination, total strangers, with other configurations, other oeconomy, and other qualities; not only the species that present themselves are new, but most of the genera, and even natural orders.

> SIR JAMES E. SMITH, *A Specimen of the Botany of New Holland*, 1793

It would appear, from the great similarity in some part or other of the different quadrupeds which we find here, that there is a promiscuous intercourse between the different sexes of all those different animals. The same observation might be made also on the fishes of the sea, on the fowls of the air, and, I may add, the trees of the forest. It was wonderful to see what a vast variety of fish were caught, which, in some part or other, partake of the shark: it is no uncommon thing to see a skait's head and shoulders to the hind part of a shark, or a shark's head to the body of a large mullet, and sometimes to the flat body of a sting-ray.

> J. HUNTER, *An Historical Journal of the Transactions at Port Jackson and Norfolk Island,* 1793

The finest terras's, lawns, and grottos, with distinct plantations of the tallest and most stately trees I ever saw in any nobleman's grounds in England, cannot excel in beauty those w'h nature now presented to our view. The singing of the various birds amongst the trees, and the flight of the numerous parraquets, lorrequets, cockatoos, and maccaws, made all around appear like in enchantment; the stupendous rocks from the summit of the hills and down to the very water's edge hang'g over in a most awful manner from above, and form'g the most commodious quays by the water, beggar'd all description.

> A. BOWES, MS journal, 1788

VIEWS OF SOUTH AMERICA

On leaving the Island del Diamante, where the Zambos, who speak Spanish, cultivate the sugar-cane, we entered into a grand and wild domain of nature. The air was filled with countless flamingoes (*Phoenicopterus*) and other water-fowl, which seemed to stand forth from the blue sky like a dark cloud in ever-varying outlines. The bed of the river had here contracted to less than 1000 feet, and formed a perfectly straight canal, which was inclosed on both sides by thick woods. The margin of the forest presents a singular spectacle. In front of the almost impenetrable wall of colossal trunks of Caesalpinia, Cedrela, and Desmanthus, there rises with the greatest regularity on the sandy bank of the river, a low hedge of Sauso, only four feet high; it consists of a small shrub, *Hermesia castanifolia*, which forms a new genus of the family of Euphorbiaceae. A few slender, thorny palms, called by the Spaniards Piritu and Corozo (perhaps species of *Martinezia* or *Bactris*) stand close alongside; the whole resembling a trimmed garden hedge, with gate-like openings at considerable distances from each other, formed undoubtedly by the large four-footed animals of the forests, for convenient access to the river. At sunset, and more particularly at break of day, the American Tiger, the Tapir, and the Peccary (*Pecari, Dicotyles*) may be seen coming forth from these openings accompanied by their young, to give them drink. When they are disturbed by a passing Indian canoe, and are about to retreat into the forest, they do not attempt to rush violently through these hedges of Sauso, but proceed deliberately along the bank, between the hedge and river, affording the traveller the gratification of watching their motions for sometimes four or five hundred paces, until they disappear through the nearest opening. During a seventy-four days' almost uninterrupted river navigation of 1420 miles up the Orinoco, to the neighborhood of its sources, and along the Cassiquiare, and the Rio Negro—during the whole of which time we were confined to a narrow canoe—the same spectacle presented itself to our view at many different points, and, I may add, always with renewed excitement. There came to drink, bathe, or fish, groups of creatures belonging to the most opposite species of animals; the larger mammalia with many-coloured herons, palamedeas with the proudly-strutting curassow (*Crax Alector, C. Pauxi*). 'It is here as in Paradise' (*es como en el Paradiso*), remarked with pious air our steersman, an old Indian, who had been brought up in the house of an ecclesiastic. But the gentle peace of the primitive golden age does not reign in the paradise of these American animals, they stand apart, watch, and avoid each other. The Capybara, a cavy (or river-hog) three or four feet long (a colossal repetition of the common Brazilian cavy, (*Cavia Aguti*), is devoured in the river by the crocodile, and on the shore by the tiger. They run so badly, that we were frequently able to overtake and capture several from among the numerous herds.

Below the mission of Santa Barbara de Arichuna we passed the night as usual in the open air, on a sandy flat, on the bank of the Apure, skirted by the impenetrable forest. We had some difficulty in finding dry wood to kindle the fires with which it is here customary to surround the bivouac, as a safeguard against the attacks of the Jaguar. The air was bland and soft, and the moon shone brightly. Several crocodiles approached the bank;

and I have observed that fire attracts these creatures as it does our crabs and many other aquatic animals. The oars of our boats were fixed upright in the ground, to support our hammocks. Deep stillness prevailed, only broken at intervals by the blowing of the fresh-water dolphins, which are peculiar to the river net-work of the Orinoco (as, according to Colebrooke, they are also to the Ganges, as high up the river as Benares); they followed each other in long tracks.

After eleven o'clock, such a noise began in the contiguous forest, that for the remainder of the night all sleep was impossible. The wild cries of animals rung through the woods. Among the many voices which resounded together, the Indians could only recognize those which, after short pauses, were heard singly. There was the monotonous, plaintive, cry of the Aluates (howling monkeys), the whining, flute-like notes of the small sapajous, the grunting murmur of the striped nocturnal ape (*Nyctipithecus trivirgatus*, which I was the first to describe), the fitful roar of the great tiger, the Cuguar or maneless American lion, the peccary, the sloth, and a host of parrots, parraquas (*Ortalides*), and other pheasant-like birds. Whenever the tigers approached the edge of the forest, our dog, who before had barked incessantly, came howling to seek protection under the hammocks. Sometimes the cry of the tiger resounded from the branches of a tree, and was then always accompanied by the plaintive piping tones of the apes, who were endeavouring to escape from the unwonted pursuit.

If one asks the Indians why such a continuous noise is heard on certain nights, they answer, with a smile, that 'the animals are rejoicing in the beautiful moonlight, and celebrating the return of the full moon.' To me the scene appeared rather to be owing to an accidental, long-continued, and gradually increasing conflict among the animals. Thus, for instance, the jaguar will pursue the peccaries and the tapirs, which, densely crowded together, burst through the barrier of tree-like shrubs which opposes their flight. Terrified at the confusion, the monkeys on the tops of the trees join their cries with those of the larger animals. This arouses the tribes of birds who build their nests in communities, and suddenly the whole animal world is in a state of commotion. Further experience taught us that the voices were loudest during violent storms of rain, or when the thunder echoed and the lightning flashed through the depths of the woods. The good-natured Franciscan monk who (notwithstanding the fever from which he had been suffering for many months), accompanied us through the cataracts of Atures and Maypures to San Carlos, on the Rio Negro, and to the Brazilian coast, used to say, when apprehensive of a storm at night, 'May Heaven grant a quiet night both to us and to the wild beasts of the forest!'

ALEXANDER VON HUMBOLDT, *Views of Nature*, 1808

THE GALAPAGOS ARCHIPELAGO

I have not as yet noticed by far the most remarkable feature in the natural history of this archipelago; it is, that the different islands to a considerable extent are inhabited by a different set of beings. My attention was first called to this fact by the Vice-Governor, Mr. Lawson, declaring that the tortoises differed from the different islands, and that he could with certainty tell from which island any one was brought. I did not for some time pay sufficient attention to this statement, and I had already partially mingled together the collections from two of the islands. I never dreamed that islands, about fifty or sixty miles apart, and most of them in sight of each other, formed of precisely the same rocks, placed under a quite similar climate, rising to a nearly equal height, would have been differently tenanted; but we shall soon see that this is the case. It is the fate of most voyagers, no sooner to discover what is most interesting in any locality, than they are hurried from it; but I ought, perhaps, to be thankful that I obtained sufficient material to establish this most remarkable fact in the distribution of organic beings.

The inhabitants, as I have said, state that they can distinguish the tortoises from the different islands; and that they differ not only in size, but in other characters. Captain Porter has described those from Charles and from the nearest island to it, namely, Hood Island, as having their shells in front thick and turned up like a Spanish saddle, whilst the tortoises from James Island are rounder, blacker, and have a better taste when cooked. M. Bibron, moreover, informs me that he has seen what he considers two distinct species of tortoise from the Galapagos, but he does not know from which islands. The specimens that I brought from three islands were young ones; and probably owing to this cause, neither Mr. Gray nor myself could find in them any specific differences. I have remarked that the marine Amblyrhynchus was larger at Albemarle Island than elsewhere; and M. Bibron informs me that he has seen two distinct aquatic species of this genus;

so that the different islands probably have their representative species or races of the Amblyrhynchus, as well as of the tortoise. My attention was first thoroughly aroused, by comparing together the numerous specimens, shot by myself and several other parties on board, of the mocking-thrushes, when, to my astonishment, I discovered that all those from Charles Island belonged to one species (Mimus trifasciatus); all from Albemarle Island to M. parvulus; and all from James and Chatham Islands (between which two other islands are situated, as connecting links) belonged to M. melanotis. These two latter species are closely allied, and would by some ornithologists be considered as only well-marked races or varieties; but the Mimus trifasciatus is very distinct. Unfortunately most of the specimens of the finch tribe were mingled together; but I have strong reasons to suspect that some of the species of the sub-group Geospiza are confined to separate islands. If the different islands have their representatives of Geospiza, it may help to explain the singularly large number of the species of this sub-group in this one small archipelago, and as a probable consequence of their numbers, the perfectly graduated series in the size of their beaks. Two species of the sub-group Cactornis, and two of Camarhynchus, were procured in the archipelago; and of the numerous specimens of these two sub-groups shot by four collectors at James Island, all were found to belong to one species of each; whereas the numerous specimens shot either on Chatham or Charles Island (for the two sets were mingled together) all belonged to the two other species: hence we may feel almost sure that these islands possess their representative species of these two sub-groups. In land-shells this law of distribution does not appear to hold good. In my very small collection of insects, Mr. Waterhouse remarks, that of those which were ticketed with their locality, not one was common to any two of the islands.

If we now turn to the Flora, we shall find the aboriginal plants of the different islands wonderfully different. I give all the following results on the high authority of my friend Dr. J. Hooker. I may premise that I indiscriminately collected everything in flower on the different islands, and fortunately kept my collections separate. Too much confidence, however, must not be placed in the proportional results, as the small collections brought home by some other naturalists, though in some respects confirming the results, plainly show that much remains to be done in the botany of this group: the Leguminosæ, moreover, have as yet been only approximately worked out:—

Name of Island	Total No. of Species	No. of Species found in other parts of the world	No. of Species confined to the Galapagos Archipelago	No. confined to the one Island	No. of Species confined to the Galapagos Archipelago, but found on more than the one Island
James Island	71	33	38	30	8
Albemarle Island	46	18	26	22	4
Chatham Island	32	16	16	12	4
Charles Island	68	39*	29	21	8

* or 29, if the probably imported plants be subtracted

Hence we have the truly wonderful fact, that in James Island, of the thirty-eight Galapageian plants, or those found in no other part of the world, thirty are exclusively confined to this one island; and in Albemarle Island, of the twenty-six aboriginal Galapageian plants, twenty-two are confined to this one island, that is, only four are at present known to grow in the other islands of the archipelago; and so on, as shown in the above table, with the plants from Chatham and Charles Islands. This fact will, perhaps, be rendered even more striking, by giving a few illustrations:— thus, Scalesia, a remarkable arborescent genus of the Compositæ, is confined to the archipelago: it has six species; one from Chatham, one from Albemarle, one from Charles Island, two from James Island, and the sixth from one of the three latter islands, but it is not known from which: not one of these six species grows on any two islands. Again, Euphorbia, a mundane or widely distributed genus, has here eight species, of which seven are confined to the archipelago, and not one found on any two islands: Acalypha and Borreria, both mundane genera, have respectively six and seven species, none of which have the same species on two islands, with the exception of one Borreria, which does occur on two islands. The species of the Compositæ are particularly local; and Dr. Hooker has furnished me with several other most striking illustrations of the difference of the species on the different islands. He remarks that this law of distribution holds good both with those genera confined to the archipelago, and those distributed in other quarters of the world: in like manner we have seen that the different islands have their proper species of the mundane genus of tortoise, and of the widely distributed American genus of the mocking-thrush, as well as of two of the Galapageian sub-groups of finches, and almost certainly of the Galapageian genus Amblyrhynchus.

The distribution of the tenants of this archipelago would not be nearly so wonderful, if, for instance, one island had a mocking-thrush, and a second island some other quite distinct genus;—if one island had its genus of lizard, and a second island another distinct genus, or none whatever;—or if the different islands were inhabited, not by representative species of the same genera of plants, but by totally different genera, as does to a certain extent hold good; for, to give one instance, a large berry-bearing tree at James Island has no representative species in Charles Island. But it is the circumstance, that several of the islands possess their own species of the tortoise, mocking-thrush, finches, and numerous plants, these species having the same general habits, occupying analogous situations, and obviously filling the same place in the natural economy of this archipelago, that strikes me with wonder. It may be suspected that some of these representative species, at least in the case of the tortoise and of some of the birds, may hereafter prove to be only well-marked races; but this would be of equally great interest to the philosophical naturalist. I have said that most of the islands are in sight of each other: I may specify that Charles Island is fifty miles from the nearest part of Chatham Island, and thirty-three miles from the nearest part of Albemarle Island. Chatham Island is sixty miles from the nearest part of James Island, but there are two intermediate islands between them which were not visited by me. James Island is only ten miles from the nearest part of Albemarle Island, but the two points where the collections were made are thirty-two miles apart. I must repeat, that neither the nature of the soil, nor height of the land, nor the climate, nor the general character of the associated beings, and therefore their action one on another, can differ much in the different islands. If there be any sensible difference in their climates, it must be between the windward group (namely Charles and Chatham Islands), and that to leeward; but there seems to be no corresponding difference in the productions of these two halves of the archipelago.

The only light which I can throw on this remarkable difference in the inhabitants of the different islands, is, that very strong currents of the sea running in a westerly and W.N.W. direction must separate, as far as transportal by the sea is concerned, the southern islands from the northern ones; and between these northern islands a strong N.W. current was observed, which must effectually separate James and Albemarle Islands. As the archipelago is free to a most remarkable degree from gales of wind, neither the birds, insects, nor lighter seeds, would be blown from island to island. And lastly, the profound depth of the ocean between the islands, and their apparently recent (in a geological sense) volcanic origin, render it highly unlikely that they were ever united; and this, probably, is a far more important consideration than any other, with respect to the geographical distribution of their inhabitants. Reviewing the facts here given, one is astonished at the amount of creative force, if such an expression may be used, displayed on these small, barren, and rocky islands; and still more so, at its diverse yet analogous action on points so near each other. I have said that the Galapagos Archipelago might be called a satellite attached to America, but it should rather be called a group of satellites, physically similar, organically distinct, yet intimately related to each other, and all related in a marked, though much lesser degree, to the great American continent.

CHARLES DARWIN, *The Voyage of the Beagle,* 1845

THE HIMALAYAS

Early this morning we proceeded upwards, our prospect more gloomy than ever. The path, which still lay up steep ridges, was very slippery, owing to the rain upon the clayey soil, and was only passable from the hold afforded by interlacing roots of trees. At 8,000 feet, some enormous detached masses of micaceous gneiss rose abruptly from the ridge, they were covered with mosses and ferns, and from their summit, 7,000 feet, a good view of the surrounding vegetation is obtained. The mass of the forest is formed of:—(1) Three species of oak, of which *Q. annulata?* with immense lamellated acorns, and leaves sixteen inches long, is the tallest and the most abundant.—(2) Chestnut.—(3) *Laurineae* of several species, all beautiful forest-trees, straight-boled, and umbrageous above.—(4) Magnolias.—(5) Arborescent rhododendrons, which commence here with the *R. arboreum.* At 8,000 and 9,000 feet, a considerable change is found in the vegetation; the gigantic purple *Magnolia campbellii* replacing the white; chestnut disappears, and several laurels; other kinds of maple are seen, with

Rhododendron argenteum, and *Stauntonia*, a handsome climber, which has beautiful pendent clusters of lilac blossoms.

At 9,000 feet we arrived on a long flat covered with lofty trees, chiefly purple magnolias, with a few oaks, great *Pyri* and two rhododendrons, thirty to forty feet high (*R. barbatum*, and *R. arboreum* var. *roseum*): *Skimmia* and *Symplocos* were the common shrubs. A beautiful orchid with purple flowers (*Coelogyne wallichii*) grew on the trunks of all the great trees, attaining a higher elevation than most other epiphytical species, for I have seen it at 10,000 feet.

A large tick infests the small bamboo, and a more hateful insect I never encountered. The traveller cannot avoid these insects coming on his person (sometimes in great numbers) as he brushes through the forest; they get inside his dress, and insert the proboscis deeply without pain. Buried head and shoulders, and retained by a barbed lancet, the tick is only to be extracted by force, which is very painful. I have devised many tortures, mechanical and chemical, to induce these disgusting intruders to withdraw the proboscis, but in vain. Leeches also swarm below 7,000 feet; a small black species above 3,000 feet, and a large yellow-brown solitary one below that elevation.

Our ascent to the summit was by the bed of a water-course, now a roaring torrent, from the heavy and incessant rain. A small *Anagallis* (like *tenella*), and a beautiful purple primrose, grew by its bank. The top of the mountain is another flat ridge, with depressions and broad pools. The number of additional species of plants found here was great, and all betokened a rapid approach to the alpine region of the Himalaya. In order of prevalence the trees were,—the scarlet *Rhododendron arboreum* and *barbatum*, as large bushy trees, both loaded with beautiful flowers and luxuriant foliage; *R. falconeri*, in point of foliage the most superb of all the Himalayan species, with trunks thirty feet high, and branches bearing at their ends only leaves eighteen inches long: these are deep green above, and covered beneath with a rich brown down. Next in abundance to these were shrubs of *Skimmia laureola*, *Symplocos*, and Hydrangea; and there were still a few purple magnolias, very large *Pyri*, like mountain ash, and the common English yew, eighteen feet in circumference, the red bark of which is used as a dye, and for staining the foreheads of Brahmins in Nepal. An erect white-flowered rose (*R. sericea*, the only species occurring in Southern Sikkim) was very abundant: its numerous inodorous flowers are pendent, apparent as a protection from the rain; and it is remarkable as being the only species having four petals instead of five . . .

We encamped amongst Rhododendrons, on a spongy soil of black vegetable matter, so oozy, that it was difficult to keep the feet dry. The rain poured in torrents all the evening, and with the calm, and the wetness of the wood, prevented our enjoying a fire. Except a transient view into Nepal, a few miles west of us, nothing was to be seen, the whole mountain being wrapped in dense masses of vapour. Gusts of wind, not felt in the forest, whistled through the gnarled and naked tree-tops; and though the temperature was 50°, this wind produced cold to the feelings.

J. D. HOOKER, *Himalayan Journals*, 1855

THE MALAY ARCHIPELAGO

I had some consolation in the birds my boys brought home daily, more especially the Paradiseas, which they at length obtained in full plumage. It was quite a relief to my mind to get these, for I could hardly have torn myself away from Aru had I not obtained specimens. But what I valued almost as much as the birds themselves was the knowledge of their habits, which I was daily obtaining both from the accounts of my hunters, and from the conversation of the natives. The birds had now commenced what the people here call their 'sácaleli,' or dancing-parties, in certain trees in the forest, which are not fruit-trees, as I at first imagined, but which have an immense head of spreading branches and large but scattered leaves, giving a clear space for the birds to play and exhibit their plumes. On one of these trees a dozen or twenty full-plumaged male birds assemble together, raise up their wings, stretch out their necks, and elevate their exquisite plumes, keeping them in a continual vibration. Between whiles they fly across from branch to branch in great excitement, so that the whole tree is filled with waving plumes in every variety of attitude and motion. The bird itself is nearly as large as a crow, and is of a rich coffee-brown color. The head and neck is of a pure straw yellow above, and rich metallic green beneath. The long plumy tufts of golden-orange feathers spring from the sides beneath each wing, and when the bird is in repose are partly concealed by them. At the time of its excitement, however, the wings are raised vertically over the back, the head is bent down and stretched out, and the long plumes are raised up and expanded till they form two magnificent golden fans, striped with deep red at the base, and fading off into the pale brown tint of the finely divided and softly waving points. The whole bird is then over-shadowed by them, the crouching body, yellow head, and emerald-green throat forming but the foundation and setting to the golden glory which waves above. When seen in this attitude, the bird of paradise really deserves its name, and must be ranked as one of the most beautiful and most wonderful of living things. I continued also to get specimens of the lovely little king-bird occa-sionally, as well as numbers of brilliant pigeons, sweet little par-roquets, and many curious small birds, most nearly resembling those of Australia and New Guinea.

ALFRED RUSSEL WALLACE, *The Malay Archipelago*, 1869

The frequent—though unthinking—callousness of much Victorian natural history writing was occasionally mitigated by a vein of black humour, as in the painter John James Audubon's (himself a great slaughterer) story of Constantine Rafinesque, and the following extract from George Donaldson's account of derring-do on the Mississippi. The violence featured in the full article is barely credible.

My wardrobe was rather limited, for with the exception of my two blankets (a red one and a blue one) I could have put the rest of it into the crown of my hat.

Two years had nearly elapsed from the time I had raised my gun and killed a '*Cedar Bird*' in the State of Massachussets, before I found myself rowing up a byou to the west of the great Mississippi. Our progress through it was necessarily slow, from the over-whelming heat of the sun; the turnings and twistings were numer-ous, from having to observe the openings and narrow passages through the prairie cane; we passed through a cedar swamp of great extent, completely inundated, the trunks of the trees, on an average, being fully eight feet under water, and some of the ani-mals which I have previously mentioned I had then an opportun-ity of seeing. The first flock of Ducks which I observed were the blue winged Teal, (*Anas discors*,) of which I shot one and wounded several. The belted Kingfisher (*Alcedo Alcyon*) was of com-mon occurrence, and would frequently perch within ten yards of me, on a drooping branch of a decayed cedar; and the familiar manner in which he appeared to recognize me, by erecting his crest and bobbing his head, was often the cause of prolonging his life. We very frequently opened into lagoons of considerable extent, and on my first entrance into one, I was deceived by what I imagined to be a black and flat bank, of about an acre in extent; this, to my confusion, on a nearer approach, was converted into a countles host of Pooldeans, (a species of Coot, *Fulica atra*,) so closely crowded together, that I was often surprised afterwards that they could find sufficient room to swim; as they permitted me to approach within fifteen yards of them, you can scarcely doubt there were some deaths and a few cripples in the collection. These birds are passed by with perfect indifference as long as the Ducks continue plentiful; and during a haze the canoe can nearly be paddled on to the top of them, which I partly did, and as I cannot show you what I killed by shooting both barrels into such a mass, I may mention that one of the men with whom I afterwards associated informed me that he killed 153 by a right and left, which I do not for one moment question.

The appearance of these birds taking wing is very picturesque; in place of raising themselves into the air, they keep tripping and spattering along the surface, supported by the flapping of their wings; and this temporary commotion produced within a still lagoon is frequently very refreshing. I think that, without fear of contradiction, I may set down the Cinereous Coot as being infinitely the most numerous of any species to be found in the swamps. We continued through a long succession of creeks and lagoons, well stocked with Ducks and Pooldeans, at which I kept loading and shooting till the sun went down beyond the prairie; and, just as he *plumped* out of sight, a common practice with him in tropical countries, we reached lake Cataahoola, a distance of fully thirty miles from the river. Our landing was a very sticky one; for, in place of getting close to the shineer, we had to get out and wade up to the knees, through mud and decayed vegetation, and carry our cargo, consisting of guns, Ducks, a quantity of rice, blankets, several bags of shot, and two jars of claret, besides other articles which some of the men had ordered. On getting to the bank, I discovered that it consisted principally of shells, which had been thrown down above the decayed reeds and prairie cane, by the Choctaw Indians, who, I was afterwards informed, had carried them there for the purpose of forming mounds, not only as places of sepulture, but also as a temple for the adoration of the Great Spirit. I very shortly afterwards gathered my blankets about me, and lay down on the shells, where I slept as soundly as ever my grandmother did on a bed of down.

The following morning, very early, I found that I had got into a new circle of friends, who were rising out of their lairs all around me; several of them were rather better sheltered than I was; there were three Americans, two Frenchmen, one Mexican, a Swede, three niggers, (runaways, I suppose,) with all of whom I became immediately acquainted. The history of these men I have no doubt would afford many a strange incident, and probably a few *dark* ones, as the swamps in that country furnish the desperate with a great city of refuge; for in such a place they are beyond all law and jurisdiction; swords, bowie knives, and pistols, are within the reach of all, and many a midnight *burial* takes place amongst the lakes. Sunday is not even known, and the chase is kept up with quite as much interest on that day as any other day; and, as the markets in the southern part of Louisiana are open on the Sabbath, the supply of Ducks and fish are expected on that day as well as on Saturday.

The confusion of tongues prevailing in the French market of New Orleans, which I afterwards visited, almost convinced me that the crowd which was dispersed at Babel had come to a focus there. The morning after my arrival, I got up out of the shells about two hours before day, and found that several of my foreign acquaintances had already *put out*. A canoe (or peroque) as they are always called, was provided for me, into which I got myself squatted, and after paddling and shoving myself through a long, zigzag, marshy, and muddy creek, quite overgrown with sword-grass, by which I got my hands severely cut, I got into a small bay, which opened into lake Catawatchaa, when I secreted myself amongst the reeds till about sun up. The Ducks then began to fly, and the shooting commenced across the lagoons, from the blinds which the men had constructed to shoot the Ducks from. The quacking of these men, in imitation of the Ducks, was so very remarkable that I never could distinguish between the one and the other; this is the great secret in Ducking, and had I not seen what they can accomplish I never would have believed it. A man concealed in a *blind* can call a flock of Ducks from an altitude of 200 yards till within fifteen feet of him, and you may then suppose how many he can kill by a well directed *right and left*. In this *accomplishment* I was found wanting; but independent of it, I had as much sport as I could desire, and many a Duck did I kill, the lustrous tints of which were little inferior to many of the Humming Birds. The Prairie Hawks were very numerous, and followed closely in the track of where so much destruction was going on. To see them stooping and hunting the wounded Ducks across the lagoons was frequently a very spirited affair, and it was nothing uncommon for one of them to alight on the back of a Duck which had just been shot, and that, too, within fifteen yards distance. The first Mallard (*Anas Boschas*) I killed, was taken possession of by one of them, but whom I quickly stretched at full length alongside the Duck, to teach him that I was quite as good a judge of Ducks, without the green peas, as himself.

In about two hours I had my peroque well loaded with both Ducks and Pooldeans. I then paddled off amongst the marshy islands, in search of Alligators, which I had no difficulty in finding. The first which I came upon were laying quite exposed, with the exception of a small portion of the tail, within the lagoon. I supposed they would turn round, and present their heads before sinking themselves; in this I was disappointed, for they *backed* into it, and immediately disappeared. I was defeated repeatedly in this way, in my attempts to bag the game; but I afterwards ran my canoe close up to them, and seldom failed to burst their heads before they could effect their escape: the average length of such as these was from seven to nine feet. Some smaller ones, which I afterwards killed, I turned belly uppermost, to make them more attractive to the Black Vultures (*Vultur atratus*) and Turkey Buzzards, which were frequent in their attendance.

There is no other way of killing Ducks within the lagoons, than from a canoe, and much care is required in shooting from it; for by doing so cross-ways, (as I explained before,) it will upset in a second, and many guns and lives have been lost in this manner. They only contain a single individual; (I refer entirely to the canoes within the swamps;) and if one should tip over, in an open lagoon with a mud bottom, the chances are, the boatman might share a similar fate with the prophet, with no hope of being vomited up again. The distances at which the men are frequently from each other, prevent the possibility of their ever being *heard*; and as for being *seen*, that is out of the question, for the tall rushes and cane-brake by which they are constantly surrounded render this impossible. It occasionally proved rather a 'coggly business for me, but I always saved my distance by about half a nose.'

On my arrival at the Shells, I found that several of the men had returned with their canoes well loaded with Ducks and Coots; with which they had cooked up a very greasy mess, better suited to the taste of an Esquimaux than of a Scotchman; this however, with some boiled rice, was very acceptable. Amongst the variety which I had killed, I found the following species: the Ruddy Duck!! (*Anas rutilans,*) the Canvass-back, (*Anas valisineria,*) the Shoveller, (*Anas clypeata,*) the Mallard, (*Anas Boschas,*) the Buffel-head, (*Anas albeola,*) the Prairie Hawk, and the Cinereous Coot. (*Fulica atra.*) Many successive days were passed in the same manner, during which I shot two other species; the Green-winged Teal, (*Anas crecca,*) and, if I mistake not, the Pintail. (*Anas glacialis.*) While laying quietly concealed in my canoe, amongst the tall reeds, I have been much delighted with the near approach and the inquiring 'peep' of the Water Rail, (*Rallus Virginianus,*) and the wonderful activity displayed by the Marsh Wren, (*Certhia palustris,*) and other Creepers *which I cannot name,* (I question if they have ever been named at all,) hunting insects up and down the long slippery canes all around me.

In shooting from a blind, the person is quite concealed; and it is only after having shot a number of Ducks that he unfastens his canoe, and picks them up from off the lagoon, where they are floating around him. He is occasionally disappointed in bagging the whole, for the Alligators now and then nab a few. I lost several Ducks without knowing how; but this was soon explained, by observing the jaws of one of these animals projected from the surface, and gobbling up a Duck, within a very short distance from me. I watched for a repetition of such a piracy; and just as I caught him gaping, I discharged a barrel right down his throat; which I am quite sure, if it did him *no harm,* did him *no good.* I have little doubt that some of the monstrous Cat-fish in these swamps, practice the same thing; and this I am inclined to believe, from their snatching at some birds in my hand, which I previously mentioned. I was astonished, on one occasion, to find a frog make an attempt to swallow a Bird called the Tyrant Fly-catcher, which I had shot. It fell into a marsh, and scarcely had it reached the water when it was seized and pulled underneath; the buoyancy, however, of the bird, of which he had only swallowed a part, raised it to the surface. I immediately shot my other barrel; which resulted in wounding the frog, which swam to the other side. Upon measurement of it afterwards, I found it to exceed sixteen inches, measuring from the extended forelegs to the extremity of the hind ones. I could mention other instances of equal voracity, and two in particular; one of a shark off the island of Porto Rico, which I fed with a quantity of shavings tied up in an old handkerchief. I gave him something else, besides the shavings, which he did not appear to relish so well. The other was that of a shark, pursuing and attacking the canoe of a nigger boy, called Isaac, while crossing a lagoon; it made two attempts, and in the last one broke several of its teeth, which it left sticking in the side of the canoe, one of which I afterwards extracted with my knife.

(To be concluded in our next.)

GEORGE DONALDSON, 'The Swamps of the Mississippi', *The Naturalist*, 1855

Of a sudden I heard a great uproar in the naturalist's room. I got up, reached the place in a few moments, and opened the door, when, to my astonishment, I saw my guest running about the room naked, holding the handle of my favourite violin, the body of which he had battered to pieces against the walls in attempting to kill the bats which had entered by the open window, probably attracted by the insects flying around his candle. I stood amazed, but he continued jumping and running round and round, until he was fairly exhausted, when he begged me to procure one of the animals for him, as he felt convinced that they belonged to 'a new species'. Although I was convinced of the contrary, I took up the bow of my demolished Cremona, and administering a smart tap to each of the bats as it came up, soon got specimens enough. The war ended, I again bade him good night, but could not help observing the state of the room. It was strewed with plants, which it would seem he had arranged into groups, but which were now scattered about in confusion. 'Never mind, Mr. Audubon,' quoth the eccentric naturalist, 'never mind, I'll soon arrange them again. I have the bats, and that's enough.'

GEORGE AUDUBON, *Ornithological Biography*, 1831–4

Before it was fairly light we lowered, and paddled as swiftly as possible to the bay where we had last seen the spout overnight. When near the spot we rested on our paddles a while, all hands looking out with intense eagerness for the first sign of the whale's appearance. There was a strange feeling among us of unlawfulness and stealth, as of ambushed pirates waiting to attack some unwary merchantman, or highwaymen waylaying a fat alderman on a country road. We spoke in whispers, for the morning was so still that a voice raised but ordinarily would have reverberated among the rocks which almost overhung us, multiplied indefinitely. A turtle rose ghost-like to the surface at my side, lifting his queer head, and, surveying us with stony gaze, vanished as silently as he came.

One looked at the other inquiringly, but the repetition of that long expiration satisfied us all that it was the placid breathing of the whale we sought somewhere close at hand. The light grew rapidly better, and we strained our eyes in every direction to discover the whereabouts of our friend, but for minutes without result. There was a ripple just audible, and away glided the mate's boat right for the near shore. Following him with our eyes, we almost immediately beheld a pale, shadowy column of white, shimmering against the dark mass of the cliff not a quarter of a mile away. Dipping our paddles with the utmost care, we made after the chief, almost holding our breath. The harpooner rose, darted once, twice, then gave a yell of triumph that ran re-echoing all around in a thousand eerie vibrations, startling the drowsy *peca* [fruit-bats] in myriads from where they hung in inverted clusters on the trees above. But, for all the notice taken by the whale, she might never have been touched. Close nestled to her side was a youngling of not more, certainly, than five days old, which sent up its baby-spout every now and then about two feet into the air. One long, wing-like fin embraced its small body, holding it close to the massive breast of the tender mother, whose only care seemed to be to protect her young, utterly regardless of her own pain and danger. If sentiment were ever permitted to interfere with such operations as ours, it might well have done so now; for while the calf continually sought to escape from the enfolding fin, making all sorts of puny struggles in the attempt, the mother scarcely moved from her position, although streaming with blood from a score of wounds. Once, indeed, as a deep-searching thrust entered her very vitals, she raised her massy flukes high in the air with an apparently involuntary movement of agony; but even in that dire time she remembered the possible danger to her young one, and laid the tremendous weapon as softly down upon the water as if it were a feather fan.

So in the most perfect quiet, with scarcely a writhe, nor any sign of flurry, she died, holding the calf to her side until her last vital spark had fled, and left it to a swift despatch with a single lance-thrust. No slaughter of a lamb ever looked more like murder. Nor, when the vast bulk and strength of the animal was considered, could a mightier example have been given of the force and quality of maternal love.

The whole business was completed in half an hour from the first sight of her, and by the mate's hand alone, none of the other boats needing to use their gear. As soon as she was dead, a hole was bored through the lips, into which a tow-line was secured, the two long fins were lashed close into the sides of the animal by an encircling line, the tips of the flukes were cut off, and away we started for the ship . . .

FRANK T. BULLEN, *The Cruise of the 'Cachalot' Round the World after Sperm Whales*, 1898

Orchids, and other glamorous exotics, were one of the bridges between the Victorian explorers and the great numbers of people in fashionable thrall to plants at home.

From the almost insufferable glare of the vertical sunshine, a few steps took me into a scene where the gloom was so sombre,—heightened doubtless by the sudden contrast,—as to cast a kind of awe over the spirit. Yet it was a beauteous gloom,—rather a subdued and softened light, like that which prevails in some old pillared cathedral when the sun's rays struggle through the many-stained glass of a painted window. Choice plants that I had been used to see fostered and tended in pots in our stove-houses at home, were there in wild and *riant* luxuriance . . . wild pines, ferns, orchids, cactuses . . . —were clustering in noble profusion of vegetable life . . .

A fine epiphyte orchid scents the air with fragrance, and it is discovered far up in the fork of some vast tree; then there is the palpitation of hope and fear as we discuss the possibility of getting it down; then come the contrivances and efforts,—pole after pole is cut and tied together with the cords which the forest-climbers afford. At length the plant is reached, and pushed off, and triumphantly bagged; but lo! while examining it, some elegant twisted shell is discovered, with its tenant snail, crawling on the leaves. Scarcely is this boxed, when a glorious butterfly rushes out of the gloom into the sunny glade, and is in a moment seen to be a novelty; then comes the excitement of pursuit . . .

PHILIP HENRY GOSSE, *The Romance of Natural History*, 1861

THE DISCOVERY OF *VICTORIA AMAZONICA*

The giant Amazonian water-lily, now known as *Victoria amazonica*, was to Queen Victoria what *Strelitzia reginae* was to Queen Charlotte. It had the right kind of mysterious, exotic pedigree necessary for a Royal plant. It was spectacularly beautiful and hugely prolific. Its brilliantly engineered floating leaves suggested analogies with our own island enterprise. And its triumphant reception at Kew—and later at the Crystal Palace, inspired by the structure of its ribbed leaves—marks the pinnacle of Victorian plant worship.

The first real news of this prodigy came to Britain in September 1837, from the young explorer Sir Robert Schomburgk, who had been sent out to British Guiana (now Guyana) by the Royal Geographical Society:

It was on the 1st of January, 1837, while contending with the difficulties that nature interposed in different forms, to stem our progress up the River Berbice (lat. 4°40′N., long. 52°W.), that we arrived at a part where the river expanded and formed a currentless basin. Some object on the southern extremity of this basin attracted my attention, and I was unable to form an idea what it could be; but, animating the crew to increase the rate of their paddling, we soon came opposite the object which had raised my curiosity, and, behold, a vegetable wonder! All calamities were forgotten; I was a botanist, and felt myself rewarded! There were gigantic leaves, five to six feet across, flat, with a broad rim, lighter green above and vivid crimson below, floating upon the water; while, in character with the wonderful foliage, I saw luxuriant flowers, each consisting of numerous petals, passing, in alternate tints, from pure white to rose and pink. The smooth water was covered with the blossoms, and as I rowed from one to the other, I always found something new to admire. The flower-stalk is an inch thick near the calyx and studded with elastic prickles, about three-quarters of an inch long. When expanded, the four-leaved calyx measures a foot in diameter, but is concealed by the expansion of the hundred-petaled corolla. This beautiful flower, when it first unfolds, is white with a pink centre; the colour spreads as the bloom increases in age; and, at a day old, the whole is rose-coloured. As if to add to the charm of this noble

Water-Lily, it diffuses a sweet scent. As in the case of others in the same tribe, the petals and stamens pass gradually into each other, and many petaloid leaves may be observed bearing vestiges of an anther. The seeds are numerous and imbedded in a spongy substance.

Ascending the river, we found this plant frequently, and the higher we advanced, the more gigantic did the specimens become; one leaf we measured was six feet five inches in diameter, and rim five inches and a half high, and the flowers a foot and a quarter across.

Schomburgk sent seeds of the water-lily to Kew but, long before any of them grew into a plant, a complex melodrama, touched with typically Victorian pomposity and moments of farce, had developed over the propriety of the species' name. Schomburgk had asked for it to be called *Nymphaea victoria*, after the Queen. But closer examination showed that it was not a *Nymphaea* at all, but a member of a quite new genus. So for a while it was *Victoria regia*—or *regina* or *regalis*, depending on which index or publication you consulted. Then the appalling discovery was made that the Queen's water-lily had already been found and described by several earlier foreign botanists. One of them, Poeppig, had given it the un-regal name of *Euryale amazonica* back in 1832 and, by the strict rules of botanic nomenclature, that was what it must be called.

Again the taxonomists had made a blunder. The water-lily was not a *Euryale* either. The rules of nomenclature now decreed that the first successful bid for each half of the name should stand, so the plant became officially *Victoria amazonica*. The Queen's name had been restored but with a somewhat embarrassing epithet in attendance. Sir William Hooker gravely pronounced that the name was 'totally unsuited to be in connection with the name of Her Most Gracious Majesty and must therefore forthwith be rejected'. Etiquette, for once, triumphed over science, and during the Queen's life the giant water-lily was referred to as *Victoria regia*.

This was merely a side-show compared to the spectacular piece of theatre in progress in the fierce light of the nation's greenhouses. Schomburgk's second batch of seeds (the first was not successful) was divided between Kew, the Duke of Northumberland's garden at Syon and the Duke of Devonshire's at Chatsworth. Each planted out their quota in conditions they thought most likely to produce the first flower. The contest rapidly turned into a race, 'as exciting in its day' as Wilfrid Blunt wrote, 'as Scott's and Amundsen's to the South Pole or the Americans' and Russians' to the moon'. In the end, Devonshire's gardener Joseph Paxton proved to have found the conditions which suited the water-lily best, and on 2 November 1849 he was able to write excitedly to the Duke (then in Ireland): 'Victoria has shown a flower!! An enormous bud like a poppy head made its appearance yesterday. It looks like a large peach placed in a cup. No words can describe the grandeur and beauty of the plant.' The Duke rushed home, and a flower was presented to the Queen at Windsor.

Kew's specimen did not come into bloom until the following summer, but proved to be a sensation, and tens of thousands of people travelled to the Gardens especially to see it. They must have felt the journey well worth while for, once it had started, *Victoria amazonica* flowered with remarkable punctuality and in grand style. At about 2 o'clock each afternoon, each new white flower bud began to emit a strong smell, compared variously to pineapples, strawberries and melons. A few hours later the petals opened and began to change colour to a rose-pink. Towards 10 o'clock they began to close. The flower's slow decline continued the following day, when the fading petals became a 'drapery of Tyrian splendour' until the flower finally sank beneath the water.

ROBERT SCHOMBURGK, *Curtis Botanical Magazine*, vol. 73, 1846; commentary by Richard Mabey, 1988

I have plucked a white lily bud just ready to expand, and, after keeping it in water for two days, have turned back its sepals with my hand and touched the lapped points of the petals, when they sprang open and rapidly expanded in my hand into a perfect blossom, with the petals as perfectly disposed at equal intervals as on their native lakes, and in this case, of course, untouched by an insect. I cut its stem short and placed it in a broad dish of water, where it sailed about under the breath of the beholder with a slight undulatory motion. The breeze of his half-suppressed admiration it was that filled its sail. It was a rare-tinted one. A kind of popular aura that may be trusted, methinks. Men will travel to the Nile to see the lotus flower, who have never seen in their glory the lotuses of their native streams.

HENRY THOREAU, *Journals*, 2 July 1852; a coincidental footnote from Thoreau on the 'popular aura' of water-lilies

A FLORAL DIAL

This representation is that of a Floral Dial, or Flower Clock, composed of English wild flowers. It is arranged on the well-known principle of the regular rest, or sleep, which is taken by plants every twenty-four hours. A flower in this wreath when shown open opposite to any hour indicates that the flower regularly opens at that time. A flower closed shows that the adjacent hour is that at which it closes. When, as from three o'clock to seven o'clock, there are two flowers placed opposite to each hour, one of the flowers indicates the morning and the other the evening hour. At eight o'clock, however, both Scarlet pimpernel and Proliferous pink tell by their opening the morning hour. Care has been exercised to choose, where it was possible, the more familiar flowers, and in each instance those which are most regular. A far brighter and more striking group might be given, but, like many an elaborately ornamented timepiece, it would not be so faithful a monitor of the passing moments as the more sober and less decorated dial. The prevailing colours of English native flowers are well shown on this wreath; yellow being most characteristic of the bloom of our islands, and next to yellow, white. Pink, though rarer, is far more common than blue; while scarlet has only two representatives in the whole country—the pimpernel and the poppy. Another characteristic feature of flowers in general is displayed on the dial. The Nottingham catchfly, and the Evening primrose—which are the flowers opening in the evening—are far more sweetly and powerfully perfumed than any of the rest. Most night-blowing flowers like these two are highly fragrant, and in this way attract those insects, which, by bearing the pollen of one blossom to the pistil of another, carry on the process of plant fertilisation. The flowers that bloom by day draw the bees and butterflies by their brilliant hues; while those that open at or after dusk, which are usually pale and colourless, are entirely dependent upon their sweet odours for attracting moths that fly by night.

NAMES OF THE FLOWERS ON THE DIAL

Yellow Goatsbeard, or Noontide Opens at III. o'Clock, a.m.
  (*Tragopodon pratense*)
Wild Succory, or Chicory
  (*Cichorium intybus*)   "   IV.   "   "

Common Nipplewort
  (*Lapsana communis*)   "   V.   "   "
Buttercup (*Ranunculus bulbosus*)   "   VI.   "   "
White Water Lily (*Nymphæa alba*)   "   VII.   "   "
Scarlet Pimpernel (*Anagallis arvensis*)   "   VIII.   "   "
Proliferous Pink (*Dianthus prolifer*)   "   VIII.   "   "
Lesser Celandine (*Ranunculus ficaria*)   "   IX.   "   "
Common Nipplewort
  (*Lapsana communis*)   Closes at   X.   "   "
Common Star of Bethlehem,
  or Lady Eleven O'clock
  (*Ornithogalum umbellatum*)   Opens at   XI.   "   "
Yellow Goatsbeard
  (*Tragopodon pratense*)   Closes at   XII.   "   "
Proliferous Pink (*Dianthus prolifer*)   "   I.   "   p.m.
Scarlet Pimpernel (*Anagallis arvensis*)   "   II.   "   "
Rough Dandelion
  (*Leontodon Hispidum*)   "   III.   "   "
Wild Succory, or Chicory
  (*Cichorium intybus*)   "   IV.   "   "
White Water Lily (*Nymphæa alba*)   "   V.   "   "
Nottingham Catchfly
  (*Silene nutans*)   Opens at   VI.   "   "
Evening Primrose (*Ænothera biennis*)   "   VII.   "   "

REVD JAMES NEIL, *Rays from the Realms of Nature, or Parables of Plant Life*, 1879

In the forests of India is found a tree of the fig tribe called the Banyan. All the branches of this tree naturally bend to the earth, and push their way downward. When they are long enough to reach the ground they take root, and soon grow into strong trunks, until in process of time the aged banyan becomes chained to earth by ten hundred ties. There is another still more remarkable tree possessing a similar property, the mangrove, which frequents the mouths of tropical rivers. Here in the slimy mud, brought down by such streams, and accumulated in loathsome marshes where they reach the sea, the mangrove flourishes. Roots are being continually sent down from its branches, for, to use the language of Mr. Kingsley's graphic description, 'every bough lowers its own living cord to take fresh hold of the foul soil below.'

The original roots of the tree, which appears as if it were built on so many piles in the water, are constantly exposed in all their web-like ramifications at each low tide, and, together with those let down in countless numbers from the branches, all seem 'one horrid complicated trap for the voyager.' The odour arising from these roots when thus laid bare causes a ceaseless curse of deadly malaria to haunt the mangrove forest. And as with these trees, so is it with the natural heart of man. All the sinner's affections and desires tend earthwards, and go out only towards the things of time and sense. If he is not early awakened these take deep root and grow, and late in life his soul is bound by a thousand ties to this perishing world, and nothing short of a miracle of divine grace can loose his fetters, or make him cease to wallow in the poisonous mire of earthly pollution.

> REVD JAMES NEIL, 'The Worldly-Minded', *Rays from the Realms of Nature, or Parables of Plant Life*, 1879

It was not in altogether a figurative sense that we regarded leaves as vegetable units, the equivalents of *zooids* in such polypidoms as a Sea-fir. Every plant, shrub, or tree may be looked upon as a colony of such units; a co-operative society on a varying scale, ranging from individuals possessed of merely a thallus, like Seaweeds, Liverworts, etc., to herbaceous plants with not more than half a dozen real leaves, and upwards to the lordly Oak and gigantic Euphorbias with their myriads. Vegetable society, like human, is therefore collected into villages, small towns, and populous cities. The leaves of a plant share in the mutual benefit of its life. A tree is a nation, with its units of leaf-population coming and going year after year. It is subject to the same laws of rise, decline, and ultimate fall, which history shows has characterised the national life of many countries. Its defensive thorns, prickles, spines, poisons, etc., are to it what such defensive resources as the army, navy, and fortifications are to a great nation. Its roots and leaves are the *trading* or wealth-accumulating members, its flowers are its expending members, its fruits the emigrating part of the population, its thorns and spines leaf-units set apart, as we do our soldiers, for the defence of the general community.

Weak animals find security in associating together, and mankind have doubtless had their own social character evolved, with all the qualities and attributes belonging to it, from adopting a similar habit. Herbivorous animals collect in herds, birds in flocks, and even insects often in dense crowds. The law which has produced such a tendency has operated in a similar manner among certain kinds of vegetation. Everybody has noticed how the Alpine Gentians, Anemones, etc., have congregated in brilliant patches in Switzerland, to the general exclusion of other plants. The Daisies and Buttercups of our English meadows, and the Poppies of our cornfields, flourish best when congregated in numbers.

A real and most important reason for such a social habit, is probably to tempt insects to their neighbourhood by exhibiting great masses of colour. Many of the 'social plants,' as Humboldt long ago termed them, have got possession of the ground through being specialised to extreme physical conditions, as the Arctic and Alpine flora are to cold, for instance. Height above the sea-level is equivalent in climate to high latitude. In our lowlands *seasonal* appearance is equivalent to both these conditions, and the earliest plants to blossom in this country belong to genera, and sometimes even species, which are Alpine, as *Chrysosplenium oppositifolium* and *C. alternifolium*; and thus we have a 'vernal' flora as well as one peculiar to the summer. All of our spring plants that are identical with Alpine or Arctic species are distinguished by flowering *earlier* in the year; in some cases as much as two or three months before their brethren within the polar circle, or on the fringe of Alpine snowfields. Consequently, we owe our beautiful and cheering 'spring flowers' to the same great physical geographical distribution which brought over the Arctic plants now found on our British mountain-tops. And they are as much acclimatised and protected by simply altering their flowering time, and blossoming earlier in the year, as if they had been left stranded on high elevations instead.

J. E. TAYLOR, 'Co-operation', *The Sagacity and Morality of Plants*, 1884

There are animals in which results so strange, fantastic, even seemingly horrible, are produced, that fallen man may be pardoned, if he shrinks from them in disgust. That, at least, must be a consequence of our own wrong state; for everything is beautiful and perfect in its place. It may be answered, 'Yes, in its place; but its place is not yours. You had no business to look at it, and must pay the penalty for intermeddling.' I doubt that answer; for surely, if man have liberty to do anything, he has liberty to search out freely his heavenly Father's works; and yet every one seems to have his antipathic animal; and I know one bred from his childhood to zoology by land and sea, and bold in asserting, and honest in feeling, that all without exception is beautiful, who yet cannot, after handling and petting and admiring all day long every uncouth and venomous beast, avoid a paroxysm of horror at the sight of the common house-spider. At all events, whether we were intruding or not, in turning this stone, we must pay a fine for having done so; for there lies an animal as foul and monstrous to the eye as 'hydra, gorgon, or chimæra dire,' and yet so wondrously fitted to its work, that we must needs endure for our own instruction to handle and to look at it. Its name, if you wish for it, is Nemertes; probably N. Borlasii; a worm of very 'low' organization, though well fitted enough for its own work. You see it? That black, shiny, knotted lump among the gravel, small enough to be taken up in a dessert spoon. Look now, as it is raised and its coils drawn out. Three feet—six— nine, at least: with a capability of seemingly endless expansion; a slimy tape of living caoutchouc, some eighth of an inch in diameter, a dark chocolate-black, with paler longitudinal lines. Is it alive? It hangs, helpless and motionless, a mere velvet string across the hand. Ask the neighbouring Annelids and the fry of the rock fishes, or put it into a vase at home, and see. It lies motionless, trailing itself among the gravel; you cannot tell where it begins or ends; it may be a dead strip of sea-weed, Himanthalia lorea, perhaps, or Chorda filum; or even a tarred string. So thinks the little fish who plays over and over it, till he touches at last what is too surely a head. In an instant a bell-shaped sucker mouth has fastened to his side. In another instant, from one lip, a concave double proboscis, just like a tapir's (another instance of the repetition of forms), has clasped him like a finger; and now begins the struggle: but in vain. He is being 'played' with such a fishing-line as the skill of a Wilson or a Stoddart never could invent; a living line, with elasticity beyond that of the most delicate fly-rod, which follows every lunge, shortening and lengthening, slipping and twining round every piece of gravel and stem of sea-weed, with a tiring drag such as no Highland wrist or step could ever bring to bear on salmon or on trout. The victim is tired now; and slowly, and yet dexterously, his blind assailant is feeling and shifting along his side, till he reaches one end of him; and then the black lips expand, and slowly and surely the curved finger begins packing him end-foremost down into the gullet, where he sinks, inch by inch, till the swelling which marks his place is lost among the coils, and he is probably macerated to a pulp long before he has reached the opposite extremity of his cave of doom. Once safe down, the black murderer slowly contracts again into a knotted heap, and lies, like a boa with a stag inside him, motionless and blest.

> CHARLES KINGSLEY, *Glaucus: or the Wonders of the Shore*, 1855. The Revd Charles Kingsley, as well as being a well-known novelist, also wrote prolifically on natural history, travel, and social questions

If anyone goes down to those shores now, if man or boy seeks to follow in our traces, let him realise at once, before he takes the trouble to roll up his sleeves, that his zeal will end in labour lost. There is nothing, now, where in our days there was so much. Then the rocks between tide and tide were submarine gardens of a beauty that seemed often to be fabulous, and was positively delusive, since, if we delicately lifted the weed-curtains of a windless pool, though we might for a moment see its sides and floor paven with living blossoms, ivory-white, rosy-red, orange and amethyst, yet all that panoply would melt away, furled into the hollow rock, if we so much as dropped a pebble in to disturb the magic dream.

Half a century ago, in many parts of the coast of Devonshire and Cornwall, where the limestone at the water's edge is wrought into crevices and hollows, the tide-line was, like Keats' Grecian vase, 'a still unravished bride of quietness'. These cups and basins were always full, whether the tide was high or low, and the only way in which they were affected was that twice in the twenty-four hours they were replenished by cold streams from the great sea, and then twice were left brimming to be vivified by the temperate movement of the upper air. They were living flower-beds, so exquisite in their perfection, that my Father, in spite of his scientific requirements, used not seldom to pause before he began to rifle them, ejaculating that it was indeed a pity to disturb such congregated beauty. The antiquity of these rock-pools, and the infinite succession of the soft and radiant

forms, sea-anemones, sea-weeds, shells, fishes, which had inhabited them, undisturbed since the creation of the world, used to
occupy my Father's fancy. We burst in, he used to say, where
no hand had ever thought of intruding before; and if the Garden
of Eden had been situate in Devonshire, Adam and Eve, stepping lightly down to bathe in the rainbow-coloured spray, would
have seen the identical sights that we now saw,—the great prawns
gliding like transparent launches, anthea waving in the twilight
its thick white waxen tentacles, and the fronds of the dulse faintly streaming on the water, like huge red banners in some reverted
atmosphere.

All this is long over, and done with. The ring of living beauty drawn about our shores was a very thin and fragile one. It had
existed all those centuries solely in consequence of the indifference, the blissful ignorance of man. These rock-basins, fringed
by corallines, filled with still water almost as pellucid as the upper
air itself, thronged with beautiful sensitive forms of life,—they
exist no longer, they are all profaned, and emptied, and vulgar-
ized. An army of 'collectors' has passed over them, and ravaged
every corner of them. The fairy paradise has been violated, the
exquisite product of centuries of natural selection has been crushed
under the rough paw of well-meaning, idle-minded curiosity. That
my Father, himself so reverent, so conservative, had by the popularity of his books acquired the direct responsibility for a calami-
ty that he had never anticipated became clear enough to himself
before many years had passed, and cost him great chagrin. No
one will see again on the shore of England what I saw in my
early childhood, the submarine vision of dark rocks, speckled and
starred with an infinite variety of colour, and streamed over by
silken flags of royal crimson and purple.

> EDMUND GOSSE, *Father and Son*, 1907. Edmund Gosse was
> the son of Philip Henry Gosse (incidentally a friend of
> Kingsley). He describes his classic autobiography, *Father and
> Son*, as 'the record of a struggle between two temperaments,
> two consciences and almost two epochs'

Weeds and Wilderness

GERARD MANLEY HOPKINS'S *famous couplet 'O let them be left, wildness and wet; / Long live the weeds and the wilderness yet' (from 'Inversnaid') was written towards the end of the Victorian period, and catches a perceptible, albeit temporary, trend towards celebrating the commonplace and the untamed. Perhaps this was simply an extension of the Victorians' desire to appropriate everything to the cause of God, Queen, and moral improvement. Yet in the second half of the century there was a clear reaction against the contemporary obsessions with the exotic and the caged, and against the fashionable emphases on competition, natural hierarchies, and the male monopoly of scientific rigour. Gentler perspectives appeared from many unexpected quarters. Mary Roberts and Anne Pratt wrote with humour and insight about native weeds. The Woolhope Naturalists' 'fungus forays' were as good an illustration of co-operative hunting strategies as Kropotkin's theory of mutual aid. And Darwin's account of the usefulness of the humble earthworm was the best possible antidote to the violent farrago some 'Darwinists' had made of his ideas.*

North America, of course, had escaped the excesses of Victorian natural history. It still possessed the vestiges of an unclaimed wilderness, and the second half of the nineteenth century was the time its spiritual virtues were discovered by the Transcendentalist school of Emerson, Thoreau, and later, John Muir. The availability of a vast, unpredictable landscape gave American nature writing a freedom of language, reference, and association which it has never lost.

The other day a member of the staff of the Lister Institute called to see me on a lousy matter, and presently drew some live Lice from his waistcoat pocket for me to see. They were contained in pill boxes with little bits of muslin stretched across the open end thro' which the Lice could thrust their little hypodermic needles when placed near the skin. He feeds them by putting these boxes into a specially constructed belt and at night ties the belt around his waist and all night sleeps in Elysium. He is not married.

In this fashion he has bred hundreds from the egg upwards and even hybridised the two different species!

In the enfranchised mind of the scientific naturalist, the usual feelings of repugnance simply do not exist. Curiosity conquers prejudice.

> W. N. P. BARBELLION, *Journal of a Disappointed Man*, 1919. W. Barbellion (the pen-name of Bruce Cummings), died of tuberculosis in 1919, aged 30. His intense and emotional journals record his battle with the disease, and the heightened perception it gave him

The young botanists eagerly availed themselves of the first opportunity for commencing their search, and their efforts did not long remain unrewarded. Fanny might be seen, with an expression of lively interest on her countenance, climbing the cliffs, penetrating the woods, and exploring the salt marshes. Her mother thanked God for the renewed health which tinged her cheek and gave elasticity to her step, and she gladly procured stronger boots and dresses of firm texture in which she might ramble and climb at her own free will without fear of detriment. She made no objection even to the large flower-press, which would have been thought too uncouth for many a less elegant drawing-room; nay, she quite loved the rough machine as a means of procuring health and interest to her daughter.

> MARGARET PLUES, *Rambles in Search of Wild Flowers*, 1863

What sees the stranger in passing by? A small and insignificant looking weed, covering the top of an old wall, or springing from interstices where the mortar has fallen out between the stones. What sees the botanist in this simple weed? An object of great interest; formed especially for the place which it is designed to fill; a memento of the care of its Creator, and not of the plant only, but of numerous winged creatures that depend upon the ripening of its seeds for their support. Winds may shake it in passing by, and heavy storms may beat upon it, but there it grows, renewed from year to year, and covering the herbless stones with a pleasant verdure. Yet not as a single plant, opening at one season or period of the day, and failing at another; each of its polished stems upholds a bud, in different stages of verdure or of decay. In some, the starry white corollas are fully exposed to the sun; in others, which also stand upright, the white petals have fallen off, and the four-sided and light green capsules appear conspicuous; in others, again, the stalk assumes a curved form, and the capsule bends towards the earth. Look at it when thus reversed, what a curious shape! what a wonderful arrangement! Growing frequently on the summit of high walls, or in places exposed to fierce winds and heavy showers, a peculiar provision is required for the protection of the seed. Observe, therefore, a small penthouse formed by the capsule, for the capsule in this plant is permanent; it may not wither and fall off like that of the poppy or corn-flower, which does not require its assistance. Thus protected, the seed-vessel continues reversed during a few days; at the end of which the stem straightens, and uplifts the seed-vessel to the influence of the sun. Here, then, another process is discoverable: the capsule splits into six small divisions at the top; through which both air and light are freely admitted to the enclosed seeds. When their active ministry is finished, and the seeds are fully ripe, the stem bends again, and empties, as from an urn, the innumerable seeds upon the earth. Thus does every single stem change its position at least four times, to suit the different stages of growth or of decay. Upright, when first the simple flower is unfolded to the light, with its tiny mirror-like petals, so arranged as to catch and to reflect every wandering sunbeam. Bending, when the perfecting of the seeds seems to require a downward position; or, perhaps, if it be allowable to hazard a conjecture, which, as regards the vegetable economy in this respect, must be conjectural, the capsule is reversed because the plant grows on dry places, where little moisture can be imbibed by the roots, in order that its vessels may draw in more copiously the heavy night-dews which descend at the season of its flowering. But when the influence of the sun is especially required to ripen the fully-formed seeds, the stem gradually straightens, and the heavy laden seed-vessel is held up to catch its beams. Thus it remains till again gradually resuming its downward position the seeds are deposited in the earth.

Who, that looks upon one of these small seeds, brown, and

rough, and thickly coated, could imagine that a plant would emerge from out of it, perfect in every part, and having a most curious and elaborate machinery, adapted to all the purposes of vegetable life? . . .

THE DAISY

The daisy, in common with all other plants, contains within her that unchanging substance, called carbon, which has never been obtained in a separate state, of which the taste, the smell, and colour are unknown. Infusible, and indestructible by the action of caloric; it can, therefore, neither be laid hold of, nor detained, when the vegetable in which it dwells, has fallen to decay; although existing, completely formed, in the tenderest blade of grass, or the smallest flower that opens to the sunbeams.

Who, in looking at the simple daisy, could discern the unalterable carbon that dwells within her? who might conjecture that when her flowers are seen no longer, and her leaves have lost their greenness, withering from off the parent stem, and seeming to be lost for ever, there would arise from out the decaying leaves, as a spirit from its earthly tenement, a gas, a vapour, which the eye may not behold, and which, either hovering around the place from whence it rose, or floating through the air, waits only for the emerging of the daisy, or of some herb or flower, from the parent earth, at the return of spring? Into these it becomes absorbed, and then again its active ministry is seen in the developing of leaves and blossoms, which are destined as the months roll on, to undergo a similar decay and renovation.

Thus are we instructed by the daisy, in common with her kindred of wood or field, to remember that one of the constituent parts of both animal and vegetable bodies remains unaltered, amid the changes and decompositions which continually take place. It follows, therefore, that though the pins of the mortal tabernacle have been pulled up, and the dust has returned to its kindred earth, from one generation to another, yet that the component parts are still unchanged, ready to enter into a new and glorious combination, whenever the fiat of Omnipotence shall call them forth. Man may query in his folly. How can the dead be raised up, with what body do they come, when not a trace of them remains? To this there is an answer, for the whole creation is filled with emblems. Invisible things, that relate especially to our present state of being, are made known by the things that are;

even the shrubs and flowers which grow beside our path-way, are faithful monitors, and, either in their decay or renovation, suggest to us thoughts of hope or consolation.

MARY ROBERTS, *Flowers of the Matin and Evensong*, 1845

[Hemlock] is the handsomest of our poisonous umbelliferous plants, for the numerous divisions into which its large leaves are cut render them peculiarly elegant, and one might think that it deserved, like the Chervil, to receive its name from the Greek words signifying the 'glory' of its 'leaves.' The usual height of the Hemlock is three or four feet, but it is sometimes much taller. The hollow shining stem, occasionally at its base measuring three inches round, is marked so conspicuously with purplish red spots as to render the plant of ready recognition. It has also, on its surface, a fine greyish powder, or bloom, and towards the top is much branched. The lower leaves are very large, of a peculiar but not unpleasing green tint, shining, and on long concave footstalks, which are marked with fine lines. The upper leaves are smaller and less divided. The umbels, which appear in June and July, are composed of numerous small white flowers, and the abundant seeds which succeed them are egg-shaped and slightly flattened. The odour of the plant while growing is faint and disagreeable, and stronger if we bruise the foliage; the root, which is about as thick as the finger, and has strong fleshy fibres, has exactly the odour of the parsnep.

The narcotic principle of the Hemlock is very similar to that of opium, and the leaves and seeds are more poisonous than the root. Different opinions exist as to the power of this part of the plant. On the one hand, we have Baron Störck's opinion, that it is very acrid and deleterious. He tells us that a small drop or two of its milky juice, being applied to the tongue, produced long continued pain and swelling in that organ, and deprived him of the power of speech. On the other, our great naturalist, John Ray, relates that two friends of his ate portions of the root without any effect; and instances are known in which the roots boiled have been eaten in our own days, and considered sweet as carrots. Dr. Graves, who has eaten them thus, says that the flavour is exactly like that of a parsnep. The power of this part of the plant, therefore, doubtless differs at different seasons, and it is probably also affected, like other poisonous vegetables, by climate.

Steven, a Russian botanist, states that the root is often eaten by Russian peasants. Pliny remarked of it: 'As for the stems and stalks, many there be who do eat it, both green and also boiled, or stewed between two platters;' but it is quite certain that any one in this country eating it thus would find it his last earthly repast . . .

The Hemlock grows throughout Europe in hedges and on waste grounds; and in many parts of the kingdom it is a common plant, especially near walls in the neighbourhood of towns. In several parts of Kent it is rare; and the author of these pages knows but one spot near Dover where it may be found. This is about two miles from the town, along the cliffs at the east, and nearly under the South Foreland lighthouse . . . A large bed of the Hemlock grows there, and the man occupied in the charge of the Marine Telegraph at that station has availed himself of its abundance to deck with its stems and branches his little cave in the cliffs. This has a sloping entrance, and all about it he has planted the hemlock, which attains there a great luxuriance, and is in summer six feet high, affording by its numerous branches a shelter alike from sun or shower. The owner of the cave, an intelligent man, has an eye for grace and beauty, and prizes the elegant foliage, taking care to preserve its verdure by cutting off the fruits as they appear; while the robustness given by an out-of-door life, by airs and sounds from the sea, have rendered his nervous system too strong to be injured by the odour. To him the faint smell gives no disgust, though he tells how a friend, an old coast-guardsman, who occasionally visits him, cautiously declines to subject himself to its influence, and seats himself on some crag at a distance, where he may see its branches wave in safety. The clusters of seeds are carefully preserved and hung up in the cave to furnish the produce of future summers; and the lonely watcher of the Hemlock-bed tells how the people from a village on the cliff come down to gather its leaves, which they use for external remedies for fomentations in case of pain.

> ANNE PRATT, 'Hemlock', *The Poisonous, Noxious and Suspected Plants of our Fields and Woods*, 1857. Anne Pratt was a grocer's daughter from Chatham, who wrote and illustrated a string of elegant, best-selling botany books. But poor health meant that she was able to do almost no fieldwork herself. Her elder sister gathered specimens for her herbarium and to provide material for drawing from

Worms have played a more important part in the history of the world than most persons would at first suppose. In almost all humid countries they are extraordinarily numerous, and for their size possess great muscular power. In many parts of England a weight of more than ten tons (10,516 kilogrammes) of dry earth annually passes through their bodies and is brought to the surface on each acre of land . . .

Worms prepare the ground in an excellent manner for the growth of fibrous-rooted plants and for seedlings of all kinds. They periodically expose the mould to the air, and sift it so that no stones larger than the particles which they can swallow are left in it. They mingle the whole intimately together, like a gardener who prepares fine soil for his choicest plants. In this state it is well fitted to retain moisture and to absorb all soluble substances, as well as for the process of nitrification. The bones of dead animals, the harder parts of insects, the shells of land-molluscs, leaves, twigs, &c., are before long all buried beneath the accumulated castings of worms, and are thus brought in a more or less decayed state within reach of the roots of plants. Worms likewise drag an infinite number of dead leaves and other parts of plants into their burrows, partly for the sake of plugging them up and partly as food.

The leaves which are dragged into the burrows as food, after being torn into the finest shreds, partially digested, and saturated with the intestinal and urinary secretions, are commingled with much earth. This earth forms the dark coloured, rich humus which almost everywhere covers the surface of the land with a fairly well-defined layer or mantle. Von Hensen placed two worms in a vessel 18 inches in diameter, which was filled with sand, on which fallen leaves were strewed; and these were soon dragged into their burrows to a depth of 3 inches. After about 6 weeks an almost uniform layer of sand, a centimeter ($\cdot$4 inch) in thickness, was converted into humus by having passed through the alimentary canals of these two worms. It is believed by some persons that worm-burrows, which often penetrate the ground almost perpendicularly to a depth of 5 or 6 feet, materially aid in its drainage; notwithstanding that the viscid castings piled over the mouths of the burrows prevent or check the rain-water directly entering them. They allow the air to penetrate deeply into the ground. They also greatly facilitate the downward passage of roots of moderate size; and these will be nourished by the humus with which the burrows are lined. Many seeds owe their germination to having been covered by castings; and others buried to a considerable depth beneath accumulated castings lie dormant, until at some future time they are accidentally uncovered and germinate.

Worms are poorly provided with sense-organs, for they cannot be said to see, although they can just distinguish between light and darkness; they are completely deaf, and have only a feeble power of smell; the sense of touch alone is well developed. They can therefore learn little about the outside world, and it is surprising that they should exhibit some skill in lining their burrows with their castings and with leaves, and in the case of some species in piling up their castings into tower-like constructions. But it is far more surprising that they should apparently exhibit some degree of intelligence instead of a mere blind instinctive impulse, in their manner of plugging up the mouths of their burrows. They act in nearly the same manner as would a man, who had to close a cylindrical tube with different kinds of leaves, petioles, triangles of paper, &c., for they commonly seize such objects by their pointed ends. But with thin objects a certain number are drawn in by their broader ends. They do not act in the same unvarying manner in all cases, as do most of the lower animals; for instance, they do not drag in leaves by their foot-stalks, unless the basal part of the blade is as narrow as the apex, or narrower than it.

When we behold a wide, turf-covered expanse, we should remember that its smoothness, on which so much of its beauty depends, is mainly due to all the inequalities having been slowly levelled by worms. It is a marvellous reflection that the whole of the superficial mould over any such expanse has passed, and will again pass, every few years through the bodies of worms. The plough is one of the most ancient and most valuable of man's inventions; but long before he existed the land was in fact regularly ploughed, and still continues to be thus ploughed by earthworms. It may be doubted whether there are many other animals which have played so important a part in the history of the world, as have these lowly organised creatures. Some other animals, however, still more lowly organised, namely corals, have done far more conspicuous work in having constructed innumerable reefs and islands in the great oceans; but these are almost confined to the tropical zones.

CHARLES DARWIN, *Vegetable Mould and Earthworms*, 1881

Down Farnborough Kent.
July 2nd 1855.

My dear Henslow,
—Now it has occurred to me that it would be an interesting way of testing the probability of sea-transportal of seeds, to make a list of all the European plants found in the Azores,—a very oceanic archipelago—collect seeds and try if they would stand a pretty long immersion.—Do you think the most able of your little girls would like to collect for me a packet of seeds of such plants as grow near Hitcham, I paying, say 3d for each packet: it would put a few shillings into their pockets and would be an ENORMOUS advantage to me, for I grudge the time to collect the seeds, more especially, as *I have to learn the plants!* The experiment seems worth trying; what do you think? Should you object to offering this reward or payment to your little girls? you would have to select the most conscientious ones, that I might not get wrong seeds . . .

My dear old Master,
Yours affectionately,
C. Darwin.

P.S. Perhaps 3d would be hardly enough; and if the number does not turn out very great it shall be 6d a packet.

*

Down 23rd [*Aug/Sept 1855?*]

My dear Henslow,
The enclosed Umbellifer has made me very unhappy: I cannot make it out: will you name it for me? I hate the whole Family. It grew 3–4 ft high in rather moist thicket. To save trouble I send envelope all ready directed.—On account of two statements made by naturalists, *viz* (one) that the most 'typical form of a species is that which produces most seed,' I am very anxious to compare number of seed of wild & cultivated plants (I can easily see how false the above aphorism is but I want precise facts) & I most curiously forget it wd. not suffice to count seeds of one umbel of Wild Celery so will you get one of your little girls to get very finest ½ wild Celery near you, & either count (& pay well for me) all the seeds, or count umbels, & count seeds in an average umbel.—I can manage Carrot & Parsnip myself, & have wild & tame plants, marked. I have got wild Cabbage & asparagus, also, in hand.—our wild Parsnips are poor, so perhaps it

wd. be good to let some little girl count. There has been another more wonderful statement made than even the above.—*viz* that rich cultivation (not merely of the individual but of the race) *lessens* the fertility of all organic beings, by which assumption several authors (as I daresay you may have noticed) have attempted to upset Malthus' most logical writing—I mention all this just to show that my odd wishes are not *absolutely* idle. Most truly yours

C. Darwin.

CHARLES DARWIN, letters to John Stevens Henslow, 1855. (Quoted in Jean Russell-Gebbett, *Henslow of Hitcham*, 1977.) John Stevens Henslow was Professor of Botany at Cambridge. He was one of Darwin's tutors, and later encouraged him to make his voyage on the *Beagle*. In 1837 he left Cambridge to become rector of Hitcham in Suffolk, where he did pioneering experiments in village community work and botanical education. Darwin repaid his old mentor by recruiting Hitcham children to help with research work

Happily enough, competition is not the rule either in the animal world or in mankind. It is limited among animals to exceptional periods, and natural selection finds better fields for its activity. Better conditions are created by the elimination of competition by means of mutual aid and mutual support. In the great struggle for life—for the greatest possible fulness and intensity of life with the least waste of energy—natural selection continually seeks out the ways precisely for avoiding competition as much as possible. The ants combine in nests and nations; they pile up their stores, they rear their cattle—and thus avoid competition; and natural selection picks out of the ants' family the species which know best how to avoid competition, with its unavoidably deleterious consequences. Most of our birds slowly move southwards as the winter comes, or gather in numberless societies and undertake long journeys—and thus avoid competition. Many rodents fall asleep when the time comes that competition should set in; while other rodents store food for the winter, and gather in large villages for obtaining the necessary protection when at work. The reindeer, when the lichens are dry in the interior of the continent, migrate towards the sea. Buffaloes cross an immense continent in order to find plenty of food. And the beavers, when they grow numerous on a river, divide into two parties, and go,

the old ones down the river, and the young ones up the river—
and avoid competition. And when animals can neither fall asleep,
nor migrate, nor lay in stores, nor themselves grow their food
like the ants, they do what the titmouse does, and what Wallace
has so charmingly described: they resort to new kinds of food—
and thus, again, avoid competition.

'Don't compete!—competition is always injurious to the species,
and you have plenty of resources to avoid it!' That is the *tendency* of nature, not always realized in full, but always present.
That is the watchword which comes to us from the bush, the
forest, the river, the ocean. 'Therefore combine—practise mutual aid! That is the surest means for giving to each and to all the
greatest safety, the best guarantee of existence and progress,
bodily, intellectual, and moral.' That is what Nature teaches us;
and that is what all those animals which have attained the highest position in their respective classes have done.

> PETER KROPOTKIN, *Mutual Aid*, 1902. The Russian Peter
> Kropotkin, best known as an anarchist philosopher, was also
> a well-travelled academic geographer

THE WOOLHOPE NATURALISTS' FIELD CLUB
THE FORAY AMONGST THE FUNGUSES, I OCT. 1869

The last meeting of the Woolhope Club of the season for the
special study of Funguses took place at Hereford, on Friday, and
was very well attended. Fungus-hunting and pheasant-shooting
interfered somewhat with each other, but on this occasion it might
not be otherwise . . .

A little before 11 o'clock the members set out for the day's
exploration. The first stoppage was made at Merryhill Common,
an excellent locality for rare funguses. A large fairy-ring, almost
complete, and fifteen yards in diameter, was observed. It was
formed by *Agaricus (Tricholoma) subpulverulentus*, and greatly to
Mr. Lees delight, several mole-hills were close to it. The ring had
slightly increased in size since last year, when it first came under
observation. Near a clump of Scotch fir trees, *Lactarius deliciosus*, was gathered, *Agaricus disseminatus*, in large patches and at
every stage of growth, and also *Gomphidius viscidus, Russula vesca,
Ag. arvensis, Ag. humilis, Hygrophorus virgineus*, and an abundance
of *Boletus granulatus*. A mole run, which had been observed when
fresh made, and which formed a considerable arc of a circle,

was then inspected. Its position could just be observed by a slightly increased freshness of the grass, but it was not occupied by funguses.

The carriages were then taken for Haywood Forest, which was to be the chief hunting ground for the day. Here it was that the rare fungus, *Strobilomyces strobiiaceus*, had been found last year. The very fact that it had been found once again in Britain had created a lively interest amongst the leading mycologists, and specimens were ardently coveted. Mr. Edwin Lees, who had been the lucky finder thereof before, was present. He had carefully marked the spot, and on alighting from his carriage made for it with all the alacrity of scientific zeal. The search however was made in vain. It was not to be found there again. Other discoveries however quickly lessened the disappointment. The curious fungus *Coprinus picaceus*, the magpie Coprinus was growing close at hand—as rare and interesting as it is venomous and ill smelling. Clusters of *Ag. fascicularis*, its equally poisonous companion *Ag. sublateritius*, and others of *Ag. melleus*, and the graceful *Ag. infundibuliformis*, attracted the attention of beginners. *Boletus luteus* was very plentiful, *Boletus scaber* was there, and some very fine specimens of its close red-capped ally *Boletus versipellis* were gathered, *Boletus piperatus*, and *B. laricinus*.

The members scattered themselves throughout the wood, and, by hallooing from time to time, kept open the communication with each other in a lively manner. A change in the energy and intonation of the call announced 'a find,' and this often happened. Dr. Bull came upon a most beautiful group of the Fly Agaric, *Agaricus muscarius*. They were seventeen in number, forming part of a ring, and were in great perfection—a sight not to be forgotten.

> 'The pillar stem, the table head,
> As with a silken carpet spread,
> Inlaid with many a brilliant dye
> Of Nature's high-wrought tapestry.'—*Bishop Mant.*

Dr. M'Cullough lighted on the interesting little *Sphærobolus stellatus*, which throws out its sporangium with wonderful power considering the smallness of its size. Mr. Worthington Smith found a small white club-shaped fungus, growing parasitically on decaying branches of the Brake, *Pteris aquilina*; an especial acquisition. Though very plentiful here, it is a rare species, founded by Berkeley under the name of *Pistillaria puberula*. The Rev. W. Houghton found *Agaricus (Pholiota) radicosus*. The Rev. J. E. Jones Machen, called attention to a fine specimen of the bright scarlet *Peziza aurantia*, and many others were soon found . . .

The carriages were rejoined, and a ride pretty and interesting throughout, was brought to a satisfactory conclusion by the arrival at the Green Dragen.

The opportunity was now taken to examine the collection of Funguses which had been brought to the meeting. Several ladies honoured the Club by bringing Funguses, and coming to see the exhibition themselves. The Edible Funguses seemed to attract the greatest interest, and they were placed in the centre of the table. There was a fine specimen of a white Truffle, *Melanogaster variegatus*, found by the Rev. W. Houghton in the Lilleshall Woods, Shropshire. A couple of specimens of the *Lycoperdon giganteum*, or Giant Puff-ball, 'the vegetable egg,' as it has been termed from its light and excellent behaviour under culinary treatment; one of them was sent by Captain Hereford from Sufton. The *Fistulina Hepatica*, or vegetable beef steak, was there; *Lactarius deliciosus*, or orange-milk Agaric, good as its name denotes; *Agaricus orcella*, or vegetable sweetbread; *Hydnum repandum*, 'good as oysters,' says Dr. Badham; *Cantharellus cibarius*, the Chanterelle; *Coprinus comatus*, the maned agaric; *Agaricus rubescens*, the brown warty agaric; *Russula alutacea*, the buff gilled sweet Russula; *Marasmius oreades*, the Champignon or Fairy-ring Fungus; *Agaricus procerus*, the Parasol agaric, and its close allies *Ag. excoriatus* and *Ag. rachodes*; an abundance of the ordinary mushroom *Agaricus campestris*, in at least three varities; and a superabundance of the *Agaricus arvensis*, the large field, or Horse Mushroom, for just at that time it was so abundant and so fine that everybody brought or sent it . . .

The dinner took place at half-past four o'clock, being absolutely rendered late by the interest excited in the exhibition of Funguses.

The turbot and cod fish were no sooner dismissed than the specialities which require notice—the Funguses—were served separately as *entrées*, not that it was most favourable to them by any means, but that the taste of the members might be more critically exercised upon them.

The following Funguses were served. The Maned Agaric, *Coprinus comatus*, called 'The Agaric of Civilization,' because it requires not the undisturbed and airy pasture essential to so many kinds, but springs up on road sides or new ground at our very doors. It was simply cooked with the ordinary condiments of butter, pepper, and salt, and served on toast.

The next was a Giant Puff-ball, *Lycoperdon giganteum*, sliced, and fried with yolk of egg and fine herbs.

The third was the Vegetable Sweatbread, *Agaricus orcella*, which with the fourth, The Champignon, *Marasmius oreades*, was served in white sauce.

It is difficult to ascertain the exact estimation of the several dishes, but it is certain that they were very generally partaken of, and the success, so far as could be judged, was very decided. Three gentlemen thought the cook had not done justice to the white-maned agaric—a satisfactory proof that they at least had discovered its merits through the account published in the Transactions of last year. Several thought the *Orcella* and the *Oreades* excellent, but the chief merit was unquestionably awarded to the Puff-ball—'How do you recognise it?—Can you be sure of it?—How was it cooked?' &c., &c., were questions asked over and over again. It was fortunate that on the table a small specimen still remained to show them how easy it was to distinguish it from every other kind of fungus.

'The Foray Amongst the Funguses', Transactions of the Woolhope Naturalists' Field Club, 1869

The old year having passed away, and a new one having dawned, I have felt disposed to imitate my good friends the birds, by shaking my wings a little;—the more readily, seeing that the unusual mildness of the season has prematurely induced a tendency towards that feeling in man and animals generally.

Winter, known as such, has not yet appeared amongst us. In its stead, however, we have had a long succession of trying weather, injurious in its effects upon the earth and most of its inhabitants. Sickness in every form, and death in its ruthless ravages, have been ever before our eyes. Seldom, indeed, has there been a greater mortality known, in a given time, amongst mankind and the lower animals. Even the very nature of the latter appears to have undergone a temporary change; for, up to December, the voices of most of our autumnal and winter songsters (in all quarters, I hear) have been all-but silent. I have noticed their general depression of spirit, times out of number. I felt that they, like ourselves, were suffering from some hidden cause common to us all.

The first magical change in this matter, in my vicinity, and for many miles around, was on the 5th of January, instant. On that morning, the moon lingered much longer than usual,—daring even to face the sun, and lovingly to dispute with him the sovereignty of the day. The feathered tribe, to my amazement and delight, took a part in this unusual phenomenon. Robins, thrushes, blackbirds, hedge-sparrows, and wrens, for the first time became really 'vocal,' and poured forth strains *by moonlight* that indeed 'waked the groves.' I had, before, imagined them diminished in numbers,—wondered whither they had fled, (if alive,)—deemed our lovely park deserted by the choir; and now all Heaven resounded with their music! I had risen at six, a.m., (my usual hour,) and was therefore present at the birds' 'early matins.'

I will not dwell upon this, beyond remarking, that the extraordinary effect I have mentioned, produced on the *physique* of the birds,—gifting them at once with a pure vocal melody, appears to have been general on this identical day. The same genial weather ruled from January 5th to January 8th, on which day, as I shall presently tell you, I too underwent a similar organic change.

I believe few persons can say, with truth, that they are in the habit of hearing the blackbird in musical voice, so early as the day I have named. There however he sat, perched up aloft, and might be heard discoursing music most melodious. I should note, too, that on and after this day our little birds exhibited all the amorous dalliance so pleasingly noticed by Thomson, in his 'Spring.' They seemed to become mated as if by magic; to have wooed, courted, won, and espoused their hearts' idols, without many of the formal 'protestations' usually resorted to on these 'interesting occasions.' Early incubation is evidently the order of the day.

Here I must leave the lover of nature to his vivid imagination, which can easily fill up the details of what I merely shadow in outline. Such a change in so short a time, from a state of apathy and sickness to one of Nature's holidays, arrayed in the pleasing charms of early Spring, (brilliantly shone upon, too, by the mighty Sol, in his increasing strength,)—may be conceived, though not expressible in words. Sickness began to wear itself out quickly. The birds felt the influence of Nature, and so did I. On Monday, Jan. 8th, I rose as usual. The metropolitan carriage called for me at eight o'clock. I was habited, and ready to start. But there was something so genial in the atmosphere, which touched my spirits, (whilst opening the garden gate to make my exit,) that I felt impelled to shake my head at the coachman. This signified that he was to go on without me. 'Out of sight, out of mind,' thought I, as I retraced my steps, determined to do something out of the common way.

Now the voices of the birds were every moment becoming more musical. It was too much for me. 'A walk,' shouted I, mentally,—'and a long one!' The air freshened, and the sun peeped out, as my mind became decided. An over-coat, weighing some eight ounces, was thrown on my shoulders; a trusty stick was my companion; and away, at once, I bounded.

> WILLIAM KIDD, 'Progress of the Seasons. The Operations of Nature—January', *The Naturalist* magazine, 1855

On January 30, 1898, I passed by the water and saw the gulls there, where indeed they have spent most of the daylight hours since the first week in October. It was a rough wild morning; the hurrying masses of dark cloud cast a gloom below that was like twilight; and though there was no mist the trees and buildings surrounding the park appeared vague and distant. The water, too, looked strange in its intense blackness, which was not hidden by the silver-grey light on the surface, for the surface was everywhere rent and broken by the wind, showing the blackness beneath. Some of the gulls—about 150 I thought—were on the water together in a close flock, tailing off to a point, all with their red beaks pointing one way to the gale. Seeing them thus, sitting high as their manner is, tossed up and down with the tumbling water, yet every bird keeping his place in the company, their whiteness and buoyancy in that dark setting was quite wonderful. It was a picture of black winter and beautiful wild bird life which would have had a rare attraction even in the desert places of the earth; in London it could not be witnessed without feelings of surprise and gratitude.

We see in this punctual return of the gulls, bringing their young with them, that a new habit has been acquired, a tradition formed, which has given to London a new and exceedingly beautiful ornament, of more value than many works of art.

> W. H. HUDSON, *Birds in London*, 1898

Bushes and short trees blown like the flame of a candle. Long-tailed tits on dead stems of willow herb 4 ft. high. Rushes, half dead, half green.

Never go for a walk in the fields without seeing one thing at least however small to give me hope, the frond of a fern among dead leaves.

RICHARD JEFFERIES, *Journals*, 8 Jan. 1884

Coming like a white wall the rain reaches me, and in an instant everything is gone from sight that is more than ten yards distant. The narrow upland road is beaten to a darker hue, and two runnels of water rush along at the sides, where, when the chalk-laden streamlets dry, blue splinters of flint will be exposed in the channels. For a moment the air seems driven away by the sudden pressure, and I catch my breath and stand still with one shoulder forward to receive the blow. Hiss, the land shudders under the cold onslaught; hiss, and on the blast goes, and the sound with it, for the very fury of the rain, after the first second, drowns its own noise. There is not a single creature visible, the low and stunted hedgerows, bare of leaf, could conceal nothing; the rain passes straight through to the ground. Crooked and gnarled, the bushes are locked together as if in no other way could they hold themselves against the gales. Such little grass as there is on the mounds is thin and short, and could not hide a mouse. There is no finch, sparrow, thrush, blackbird. As the wave of rain passes over and leaves a hollow between the waters, that which has gone and that to come, the ploughed lands on either side are seen to be equally bare. In furrows full of water, a hare would not sit, nor partridge run; the larks, the patient larks which endure almost everything, even they have gone. Furrow on furrow with flints dotted on their slopes, and chalk lumps, that is all. The cold earth gives no sweet petal of flower, nor can any bud of thought or bloom of imagination start forth in the mind. But step by step, forcing a way through the rain and over the ridge, I find a small and stunted copse down in the next hollow. It is rather a wide hedge than a copse, and stands by the road in the corner of a field. The boughs are bare; still they break the storm, and it is a relief to wait a while there and rest. After a minute or so the eye gets accustomed to the branches and finds a line of sight through the narrow end of the copse. Within twenty yards—just outside the copse—there are a number of lapwings, dispersed

about the furrows. One runs a few feet forward and picks something from the ground; another runs in the same manner to one
side; a third rushes in still a third direction. Their crests, their
green-tinted wings, and white breasts are not disarranged by the
torrent. Something in the style of the birds recalls the wagtail,
though they are so much larger. Beyond these are half a dozen
more, and in a straggling line others extend out into the field.
They have found some slight shelter here from the sweeping of
the rain and wind, and are not obliged to face it as in the open.
Minutely searching every clod they gather their food in imperceptible items from the surface.

Sodden leaves lie in the furrows along the side of the copse;
broken and decaying burdocks still uphold their jagged stems,
but will be soaked away by degrees; dank grasses droop outwards;
the red seed of a dock is all that remains of the berries and fruit,
the seeds and grain of autumn. Like the hedge, the copse is
vacant. Nothing moves within, watch as carefully as I may. The
boughs are blackened by wet and would touch cold. From the
grasses to the branches there is nothing any one would like to
handle, and I stand apart even from the bush that keeps away
the rain. The green plovers are the only things of life that save
the earth from utter loneliness. Heavily as the rain may fall, cold
as the saturated wind may blow, the plovers remind us of the
beauty of shape, colour, and animation. They seem too slender to
withstand the blast—they should have gone with the swallows—
too delicate for these rude hours; yet they alone face them.

Once more the wave of rain has passed, and yonder the hills
appear; these are but uplands. The nearest and highest has a
green rampart, visible for a moment against the dark sky, and
then again wrapped in a toga of misty cloud. So the chilled Roman
drew his toga around him in ancient days as from that spot he
looked wistfully southwards and thought of Italy. Wee-ah-wee!
Some chance movement has been noticed by the nearest bird,
and away they go at once as if with the same wings, sweeping
overhead, then to the right, then to the left, and then back again,
till at last lost in the coming shower. After they have thus vibrat-
ed to and fro long enough, like a pendulum coming to rest, they
will alight in the open field on the ridge behind. There in drilled
ranks, well closed together, all facing the same way, they will
stand for hours. Let us go also and let the shower conceal them.
Another time my path leads over the hills.

It is afternoon, which in winter is evening. The sward of the down is dry under foot, but hard, and does not lift the instep with the springy feel of summer. The sky is gone, it is not clouded, it is swathed in gloom. Upwards the still air thickens, and there is no arch or vault of heaven. Formless and vague, it seems some vast shadow descending. The sun has disappeared, and the light there still is, is left in the atmosphere enclosed by the gloomy mist as pools are left by a receding tide. Through the sand the water slips, and through the mist the light glides away. Nearer comes the formless shadow, and the visible earth grows smaller. The path has faded, and there are no means on the open downs of knowing whether the direction pursued is right or wrong, till a boulder (which is a landmark) is perceived. Thence the way is down the slope, the last and limit of the hills there. It is a rough descent, the paths worn by sheep may at any moment cause a stumble. At the foot is a waggon-track beside a low hedge, enclosing the first arable field. The hedge is a guide, but the ruts are deep, and it still needs slow and careful walking. Wee-ah-wee! Up from the dusky surface of the arable field springs a plover, and the notes are immediately repeated by another. They can just be seen as darker bodies against the shadow as they fly overhead. Wee-ah-wee! The sound grows fainter as they fetch a longer circle in the gloom.

There is another winter resort of plovers in the valley where a barren waste was ploughed some years ago. A few furze bushes still stand in the hedges about it, and the corners are full of rushes. Not all the grubbing of furze and bushes, the deep ploughing and draining, has succeeded in rendering the place fertile like the adjacent fields. The character of a marsh adheres to it still. So long as there is a crop, the lapwings keep away, but as soon as the ploughs turn up the ground in autumn they return. The place lies low, and level with the waters in the ponds and streamlets. A mist hangs about it in the evening, and even when there is none, there is a distinct difference in the atmosphere while passing it. From their hereditary home the lapwings cannot be entirely driven away. Out of the mist comes their plaintive cry; they are hidden, and their exact locality is not to be discovered. Where winter rules most ruthlessly, where darkness is deepest in daylight, there the slender plovers stay undaunted.

RICHARD JEFFERIES, 'Haunts of the Lapwing', *Good Words*, January 1883; first collected in *The Open Air*, 1885

The frost held for many weeks, until the birds were dying rapidly. Everywhere in the fields and under the hedges lay the ragged remains of lapwings, starlings, thrushes, redwings, innumerable ragged, bloody cloaks of birds, whence the flesh was eaten by invisible beasts of prey.

Then, quite suddenly, one morning, the change came. The wind went to the south, came off the sea warm and soothing. In the afternoon there were little gleams of sunshine, and the doves began, without interval, slowly and awkwardly to coo. The doves were cooing, though with a laboured sound, as if they were still winter-stunned. Nevertheless, all the afternoon they continued their noise, in the mild air, before the frost had thawed off the road. At evening the wind blew gently, still gathering a bruising quality of frost from the hard earth. Then, in the yellow-gleamy sunset, wild birds began to whistle faintly in the blackthorn thickets of the stream-bottom.

It was startling and almost frightening, after the heavy silence of frost. How could they sing at once, when the ground was thickly strewn with the torn carcases of birds? Yet out of the evening came the uncertain, silvery sounds that made one's soul start alert, almost with fear. How could the little silver bugles sound the rally so swiftly, in the soft air, when the earth was yet bound? Yet the birds continued their whistling, rather dimly and brokenly, but throwing the threads of silver, germinating noise into the air.

It was almost a pain to realize, so swiftly, the new world. *Le monde est mort. Vive le monde!* But the birds omitted even the first part of the announcement, their cry was only a faint, blind, fecund '*vive!*'

There is another world. The winter is gone. There is a new world of spring. The voice of the turtle is heard in the land. But the flesh shrinks from so sudden a transition. Surely the call is premature, while the clods are still frozen, and the ground is littered with the remains of wings! Yet we have no choice. In the bottoms of impenetrable blackthorn, each evening and morning now, out flickers a whistling of birds.

Where does it come from, the song? After so long a cruelty, how can they make it up so quickly? But it bubbles through them, they are like little well-heads, little fountain-heads whence the spring trickles and bubbles forth. It is not of their own doing. In their throats the new life distils itself into sound. It is the rising of the silvery sap of a new summer, gurgling itself forth.

All the time, whilst the earth lay choked and killed and winter-mortified, the deep undersprings were quiet. They only wait for the ponderous encumbrance of the old order to give way, yield in the thaw, and there they are, a silver realm at once. Under the surge of ruin, unmitigated winter, lies the silver potentiality of all blossom. One day the black tide must spend itself and fade back. Then all-suddenly appears the crocus, hovering triumphant in the rear, and we know the order has changed, there is a new régime, sound of a new *Vive! vive!*

> D. H. LAWRENCE, 'Whistling of Birds', *Athenaeum*, April 1919

It was a beautiful day here to-day, with bright, new, wide-opened sunshine, and lovely new scents in the fresh air, as if new blood were rising. And the sea came in great long waves thundering splendidly from the unknown. It is perfect, with a strong, pure wind blowing. What does it matter about that seething scrim-mage of mankind in Europe? If that were indeed the only truth, one might indeed despair.

> D. H. LAWRENCE, *Journals*, 7 Feb. 1916

I have been scribbling on till it is not far from midnight; but I cannot put down my pen without making yet one more note. Yesterday, April 16th, is the day on which the Nightingale is generally heard for the first time in this part of Hertfordshire. I recollected just now that I had omitted to listen for it; so, to remedy my error as far as possible, I laid down my pen, and soft-ly unbarred the front-door, for all the household but myself were asleep. A charming calm night, a bright moon, clear starlight, no sound but the distant rumbling of a railway train: it dies away: out of its ruins rises a faint shrill piping, indicating pain rather than rejoicing; and before that is well ended, out bursts the li-quid gurgling note that no instrument but the throat of the Nightingale can produce. The Nightingale is arrived, and, happy augury, I have heard his song before that of the Cuckoo.

> REVD C. A. JOHNS, *Home Walks and Holiday Rambles*, 1863. Revd Johns was best known as the author of *Flowers of the Field*, 1853

In the old days of the New England transcendentalists, it used to be stated that the cosmos was a reflection of man, that his shadow ran a long way out through nature. Though the idea may be, in some sense, out of fashion, I would venture to remark that men like Emerson and Thoreau, whose interior thoughts contained a place for muskrats, bean fields, and uninhabited peaks, were closer to an analysis of man's original nature, his soul, if you will, than much that has gone on in laboratories since. A wilderness exists in man which refuses to be studied. 'There has been but the sun and the eye since the beginning,' Thoreau once wrote, and some of us prefer to have that eye round, open, and as undomesticated as an owl's in a primeval forest—a forest which invisibly surrounds us still.

> LOREN EISELEY, *Notebooks*, *c*.1960; from *The Lost Notebooks of Loren Eiseley*, ed. Kenneth Heuer, 1987

Jan. 3. Down railroad to Lincoln Bridge.
The evergreens appear to relieve themselves soonest of the ice, perhaps because of the reflection from their leaves. Those trees, like the maples and hickories, which have most spray and branches make the finest show of ice. This afternoon it snows, the snow lodging on the ice, which still adheres to the trees. The more completely the trees are changed to ice trees, to spirits of trees, the finer. Instead of the minute frost-work on a window, you have whole forests of silver boughs. I refer to the last two days. The 'brattling' of the ice. Is not that the word? Along some causeway or fence in the meadow, the trees are changed into silvery wisps. Nothing dark met the eye, but a silvery sheen, precisely as if the whole tree—trunk, boughs, and twigs—were converted into burnished silver. You exclaimed at every hedgerow. Sometimes a clump of birches fell over every way in graceful ostrich-plumes, all raying from one centre. You clambered over them like an ant in the grass. Then the beautifully checkered ice in the ruts, where the water had been soaked up, surpassing the richest tracery of watch-crystals! Suddenly all is converted to crystal. The world is a crystal palace. The trees, stiff and drooping and encased in ice, looked as if they were sculptured in marble, especially the evergreens.

I love Nature partly *because* she is not man, but a retreat from him. None of his institutions control or pervade her. There a different kind of right prevails. In her midst I can be glad with an entire gladness. If this world were all man, I could not stretch myself, I should lose all hope. He is constraint, she is freedom to me. He makes me wish for another world. She makes me content with this. None of the joys she supplies is subject to his rules and definitions. What he touches he taints. In thought he moralizes. One would think that no free, joyful labor was possible to him. How infinite and pure the least pleasure of which Nature is basis, compared with the congratulation of mankind! The joy which Nature yields is like [that] afforded by the frank words of one we love.

> Man, man is the devil,
> The source of all evil.

Methinks that these prosers, with their saws and their laws, do not know how glad a man can be. What wisdom, what warning, can prevail against gladness? There is no law so strong which a little gladness may not transgress. I have a room all to myself; it is nature. It is a place beyond the jurisdiction of human governments. Pile up your books, the records of sadness, your saws and your laws. Nature is glad outside, and her merry worms within will ere long topple them down. There is a prairie beyond your laws. Nature is a prairie for outlaws. There are two worlds, the post-office and nature. I know them both. I continually forget mankind and their institutions, as I do a bank.

HENRY THOREAU, Journal, 3 Jan. 1853

The backwoodsman and the beaver alike fell trees; the man that he may convert the forest into an olive grove that will mature its fruit only for a succeeding generation, the beaver that he may feed upon their bark or use them in the construction of his habitation. Human differs from brute action, too, in its influence upon the material world, because it is not controlled by natural compensations and balances. Natural arrangements, once disturbed by man, are not restored until he retires from the field, and leaves free scope to spontaneous recuperative energies; the wounds he inflicts upon the material creation are not healed until he withdraws the arm that gave the blow. On the other hand, I am not aware of any evidence that wild animals have ever destroyed the smallest forest, extirpated any organic species, or modified its natural character, occasioned any permanent change of terrestrial surface, or produced any disturbance of physical conditions which nature has not, of herself, repaired without the expulsion of the animal that had caused it.

GEORGE PERKINS MARSH, Man and Nature, 1864

The Great Central Plain of California, during the months of March, April, and May, was one smooth, continuous bed of honey-bloom, so marvelously rich that, in walking from one end of it to the other, a distance of more than 400 miles, your foot would press about a hundred flowers at every step. Mints, gilias, nemophilas, castilleias, and innumerable compositæ were so crowded together that, had ninety-nine per cent. Of them been taken away, the plain would still have seemed to any but Californians extravagantly flowery. The radiant, honeyful corollas, touching and overlapping, and rising above one another, glowed in the living light like a sunset sky—one sheet of purple and gold, with the bright Sacramento pouring through the midst of it from the north, the San Joaquin from the south, and their many tributaries sweeping in at right angles from the mountains, dividing the plain into sections fringed with trees.

Along the rivers there is a strip of bottom-land, countersunk beneath the general level, and wider toward the foot-hills, where magnificent oaks, from three to eight feet in diameter, cast grateful masses of shade over the open, prairie-like levels. And close along the water's edge there was a fine jungle of tropical luxuriance, composed of wild-rose and bramble bushes and a great variety of climbing vines, wreathing and interlacing the branches and trunks of willows and alders, and swinging across from summit to summit in heavy festoons. Here the wild bees reveled in fresh bloom long after the flowers of the drier plain had withered and gone to seed. And in midsummer, when the 'blackberries' were ripe, the Indians came from the mountains to feast—men, women, and babies in long, noisy trains, often joined by the farmers of the neighborhood, who gathered this wild fruit with commendable appreciation of its superior flavor, while their home orchards were full of ripe peaches, apricots, nectarines, and figs, and their vineyards were laden with grapes. But, though these luxuriant, shaggy river-beds were thus distinct from the smooth, treeless plain, they made no heavy dividing lines in general views. The whole appeared as one continuous sheet of bloom bounded only by the mountains.

When I first saw this central garden, the most extensive and regular of all the bee-pastures of the State, it seemed all one sheet of plant gold, hazy and vanishing in the distance, distinct as a new map along the foot-hills of my feet.

Descending the eastern slopes of the Coast Range through beds of gilias and lupines, and around many a breezy hillock and bush-crowned headland, I at length waded out into the midst of it. All the ground was covered, not with grass and green leaves, but with radiant corollas, about ankle-deep next the foot-hills, knee-deep or more five or six miles out. Here were bahia, madia, madaria, burrielia, chrysopsis, corethrogyne, grindelia, etc., growing in close social congregations of various shades of yellow, blending finely with the purples of clarkia, orthocarpus, and ænothera, whose delicate petals were drinking the vital sunbeams without giving back any sparkling glow.

Because so long a period of extreme drought succeeds the rainy season, most of the vegetation is composed of annuals, which spring up simultaneously, and bloom together at about the same height above the ground, the general surface being but slightly ruffled by the taller phacelias, pentstemons, and groups of *Salvia carduacea*, the king of the mints.

Sauntering in any direction, hundreds of these happy sun-plants brushed against my feet at every step and closed over them as if I were wading in liquid gold. The air was sweet with fragrance, the larks sang their blessed songs, rising on the wing as I advanced, then sinking out of sight in the polleny sod, while myriads of wild bees stirred the lower air with their monotonous hum—monotonous, yet forever fresh and sweet as every-day sunshine. Hares and spermophiles showed themselves in considerable numbers in shallow places, and small bands of antelopes were almost constantly in sight, gazing curiously from some slight elevation, and then bounding swiftly away with unrivaled grace of motion. Yet I could discover no crushed flowers to mark their track, nor, indeed, any destructive action of any wild foot or tooth whatever.

The great yellow days circled by uncounted, while I drifted toward the north, observing the countless forms of life thronging about me, lying down almost anywhere on the approach of night. And what glorious botanical beds I had! Oftentimes on awaking

I would find several new species leaning over me and looking me full in the face, so that my studies would begin before rising.

About the first of May I turned eastward, crossing the San Joaquin River between the mouths of the Tuolumne and Merced, and by the time I had reached the Sierra foot-hills most of the vegetation had gone to seed and become as dry as hay.

All the seasons of the great plain are warm or temperate, and bee-flowers are never wholly wanting; but the grand springtime— the annual resurrection—is governed by the rains, which usually set in about the middle of November or the beginning of December. Then the seeds, that for six months have lain on the ground dry and fresh as if they had been gathered into barns, at once unfold their treasured life. The general brown and purple of the ground, and the dead vegetation of the preceding year, give place to the green of mosses and liverworts and myriads of young leaves. Then one species after another comes into flower, gradually overspreading the green with yellow and purple, which lasts until May.

> JOHN MUIR, 'The Bee Pastures', *The Mountains of California*, 1894

July 27. Up and away to Lake Tenaya,—another big day, enough for a lifetime. The rocks, the air, everything speaking with audible voice or silent; joyful, wonderful, enchanting, banishing weariness and sense of time. No longing for anything now or hereafter as we go home into the mountain's heart. The level sunbeams are touching the fir tops, every leaf shining with dew. Am holding an easterly course, the deep canyon of Tenaya Creek on the right hand, Mount Hoffman on the left, and the lake straight ahead about ten miles distant, the summit of Mount Hoffman about three thousand feet above me, Tenaya Creek four thousand feet below and separated from the shallow, irregular valley, along which most of the way lies, by smooth domes and wave-ridges. Many mossy emerald bogs, meadows and gardens in rocky hollows to wade and saunter through—and what fine plants they give me, what joyful streams I have to cross, and how many views are displayed of the Hoffman and Cathedral Peak masonry, and what a wondrous breadth of shining granite pavement to walk over for the first time about the shores of the lake! On I sauntered in freedom complete; body without weight as far as I was aware; now wading through starry parnassia bogs, now through gardens shoulder deep in larkspur and lilies, grasses and rushes, shaking off showers of dew; crossing piles of crystalline moraine boulders, bright mirror pavements, and cool, cheery streams going to Yosemite; crossing bryanthus carpets and the scoured pathways of avalanches, and thickets of snow-pressed ceanothus; then down a broad, majestic stairway into the ice-sculptured lake basin.

The snow on the high mountains is melting fast, and the streams are singing bank-full, swaying softly through the level meadows and bogs, quivering with sun-spangles, swirling in potholes, resting in deep pools, leaping, shouting in wild, exulting energy over rough boulder dams, joyful, beautiful in all their forms. No Sierra landscape that I have seen holds anything truly dead or dull, or any trace of what in manufactories is called rubbish or waste; everything is perfectly clean and pure and full of divine lessons. This quick, inevitable interest attaching to everything seems marvellous until the hand of God becomes visible; then it seems reasonable that what interests Him may well interest us. When we try to pick out anything by itself, we find it hitched to everything else in the universe. One fancies a heart like our own must be beating in every crystal and cell, and we feel like stopping to speak to the plants and animals as friendly fellow mountaineers. Nature as a poet, an enthusiastic workingman, becomes more and more visible the farther and higher we go; for the mountains are fountains—beginning places, however related to sources beyond mortal ken.

JOHN MUIR, *My First Summer in the Sierras*, 1911

Cliffs and bands of rock jutted from the trees and sometimes the woods opened to make way for landslides and tumbled boulders and fans of scree. There was the scent of pine-needles and decay. Old trunks had rotted and fallen and the pale leaves of the saplings which replaced them scattered the underworld with various light and broke it into hundreds of thin sunbeams. The ghost of a track, perhaps only used by wild animals, advanced with hesitation; the matted carpet of leaves, cones, pine-needles, acorns, oak-apples, beech-mast and the split caskets of chestnuts must have been piling up forever. A tall pine had collapsed in a tangle of creepers and I was scrambling on all fours through the foxgloves and bracken underneath when my hand closed on something half-buried in leaves. It was a five-pointed stag's antler: a marvellous object, from the frilled coronet at the base to the tips of sharp tines as hard as ivory. How could something gnarled with these ancient-looking wrinkles have such a swift growth and so brief a life? They prick through a stag's brow in spring like twin thoughts breaking out of the skull, then shoot and ramify with the fluid motion of plants, fossilising as they grow; larger each year, more fiercely spiked, then scabbarded in velvet to be torn to shreds against boles and branches until the buck they have armed is ready to clear the woods of rivals; only to fall off again at the end of winter, like moulting feathers. This one was about a foot and a half long and perfectly balanced and I set off through the bracken feeling like Herne the Hunter. It was impossible to leave it there, even if I couldn't take it all the way to Constantinople.

Soon I came on four does, each with a fawn grazing or pulling at the branches that hemmed the clearing. I must have been down-wind; they only looked up when I was fairly close. They turned in a flurry, heading for the underbrush and sailing downhill in great arcs until all their white rumps had vanished in turn; and, as they took flight, a russet stag, unseen till then, looked up with a sweep of horn that was spread far wider than the antler in my hand; and while the does were curvetting past, his antlers swung out of profile into full face like a ritual separation of twin candelabra. His wide eyes were severe but unfocused, white flecks scattered the back of his tawny coat, and his hooves were neat and shining. Turning aside, he took one or two sedate and strutting paces, trotted a few more with his head and its scaffolding well back, and leaped down the slope after the does. The load of horn rose and sank with each bound; then he flew headlong through a screen of branches like a horse through a hoop and the boughs closed behind him as he crashed downhill and out of earshot.

I could hardly believe they had all been there a few seconds before. Could my antler have once been his, shed a few years back? Perhaps even now he had not reached full span, although August was beginning: I had seen no tatters of velvet . . . Anyway, the trove in my hand could just as easily have been centuries old.

Bit by bit, the shoulders of bare rock began to grudge foothold to the taller trees and I was advancing through dwarf fir and a slag-like scree covered with a spectral confusion of thistles. A pale ridge of mineral had sprung up to the right; a much loftier upheaval soared to the left, with another far away beyond it, wrinkled, ashen and shadowless, like an emanation of the noon's glare. I was moving along an empty valley of pale rocks and boulders, cheerlessly plumed here and there with little fir trees, and eventually these too died away. The warp of the mountains had led me astray. I was not sure that I was where I thought I was, or where I ought to be. It was a bleak place with the pallor of a bone-yard and a wind blowing up made it bleaker still . . .

Soon after setting off in the morning, I halted on a grassy bluff to tie up a lace when I heard a sound which was half a creak and half a ruffle. Looking over the ledge to a similar jut fifteen yards below, I found myself peering at the hunched shoulders of a very large bird at the point where his tawny feathers met plumage of a paler chestnut hue: they thatched his scalp and the nape of his neck and he was tidying up the feathers on his breast and shoulders with an imperiously curved beak. A short hop shifted the bird farther along its ledge and it was only when, with a creak, he flung out his left wing to its full stretch and began searching his armpit, that I took in his enormous size. He was close enough for every detail to show: the buff plus-four feathers covering three-quarters of his scaly legs, the yellow and black on his talons, the square-ended tail-feathers, the yellow strip at the base of his upper beak. Shifting from his armpit to his flight-feathers, he set about preening and sorting as though the night had tousled them. He folded the wing back without haste, then flung out the other in a movement which seemed to put him off balance for a moment, and continued his grooming with the same deliberation.

Careful not to move an eyelash, I must have watched for a full twenty minutes. When both wings were folded, he sat peering masterfully about, shrugging and hunching his shoulders from time to time, half-spreading a wing then folding it back, and once stretching the jaws of his beak wide in a gesture like a yawn, until at length on a sudden impulse, with a creak and a shudder, he opened both wings to their full tremendous span, rocking for a moment as though his balance were in peril; then, with another two or three hops and a slow springing movement of his plus-foured legs, he was in the air, all his flight-feathers fanning out separately and lifting at the tips as he moved his wings down, then dipping with the following upward sweep. After a few strokes, both wings came to rest and formed a single line, with all his flight feathers curling upwards again as he allowed an invisible air-current to carry him out and down and away, correcting his balance with hardly perceptible movements as he sailed out over the great gulf. A few moments later, loud but invisible flaps sounded the other side of a buttress and a second great bird followed him almost without a sound. They swayed gently, with a wide space of air between them, like ships in a mild swell. Then as they crossed the hypotenuse of shadow which stretched from the Carpathian skyline to the flanks of the Banat mountains, the morning light caught and burnished their wings and revealed them both in their proper majesty. To look down on this king and queen of birds, floating there in aloof companionship, brought a long moment of exaltation. To think the Khirgiz used golden eagles for hunting! They carried them on horseback, a seemingly impossible feat, then unhooded them over the steppe to soar and spy out antelopes and foxes and wolves and then stoop on their quarry. Hereabouts, Radu had conveyed, they sometimes rivalled wolves in decimating flocks and, I learned later on, in wreaking havoc among the sheep and goats of the Sarakatsan nomads of the Rhodope mountains, and the flocks of Radu's relations, the Koutzovlachs of the Pindus. They circle above the folds, hover, take aim, then fall like javelins and carry lambs piteously bleating into the sky . . .

I could follow their motionless hover and their languid circlings for a long time as I headed south. The encounter, within twenty-four hours of that brief Altdorfer-vision of the stag, was almost too much to take in. I wondered how near to wild boars my path had gone, or might go; and to wolves and bears. They, too, were said to keep out of men's way at this time of the year. I hadn't seen any of them; but perhaps they had seen me as I crashed past. What about the famous passion of bears for honey and the bee-hives of those harvesters? I longed to catch a glimpse of one of them ambling bandy-legged across the middle distance or reaching on tip-toe, plagued by bees, into a hollow tree after a comb. There had been movements like an unquiet spirit in the branches during the night; larger than a squirrel, it had sounded: could it have been a wild cat or a lynx? Perhaps a pine-marten.

Starting at dawn, ending at dark and only separated by light sleep, each day in the mountains seemed to contain a longer sequence of phases than a week at ground level. Twenty-four hours would spin themselves into a lifetime, and thin mountain air, sharpened faculties, the piling-up of detail and a kaleidoscope of scene-changes seemed to turn the concatenation into a kind of eternity. I felt deeply involved in these dizzy solitudes, more reluctant each minute to come down again and ready to go on forever. Thank heavens, I thought, climbing along a dark canyon of pines, no likelihood of it ending yet. But suddenly, very faintly and a long way off, there was the sound of an axe falling; then two or three. However far away, the sound struck a baleful note; it spoke of people from the lower world and the two days' solitude since leaving the shepherds had installed feelings of unchallenged ownership of everything within sight or hearing.

> PATRICK LEIGH FERMOR, *Between the Woods and the Water*, 1986: from the retrospective account of his walk from the Hook of Holland to Constantinople in the 1930s, on which he caught glimpses of the last of wild Europe

There is a great deal of talk these days about saving the environment. We must, for the environment sustains our bodies. But as humans we also require support for our spirits, and this is what certain kinds of places provide. The catalyst that converts any physical location—any environment if you will—into a place, is the process of experiencing deeply. A place is a piece of the whole environment that has been claimed by feelings. Viewed simply as a life-support system, the earth is an environment. Viewed as a resource that sustains our humanity, the earth is a collection of places. We never speak, for example, of an environment we have known; it is always places we have known—and recall. We are homesick for places, we are reminded of places, it is the sounds and smells and sights of places which haunts us and against which we often measure our present.

> ALAN GUSSOW, *A Sense of Place*, 1971

I have come here to watch muskoxen. The muskox, along with the American bison, is one of the few large animals to have survived the ice ages in North America. Most all of its companions —the mammoth, the dire wolf, the North American camel, the

short-faced bear—are extinct. The muskox abides, conspicuously alone and entirely at ease on the tundra, completely adapted
to a polar existence.

I am sitting at the edge of a precipitous bluff, several hundred
feet above the Thomsen, with a pair of high-powered binoculars.
At this point the river curves across a broad plain of seepage
meadows and tundra in a sweeping oxbow to my left; on my
right Baker Creek has cut a steep-banked gash into the badlands
to the west. Far to the south, in front of me, are clusters of black
dots; at a distance of more than two miles they arrest even the
naked eye with a strange, faint reflection. The mind knows by
that slow drift of dark points over a field of tan grasses on an
open hillside that there is life out there. But an older, deeper
mind is also alerted by the flash of light from those distant, long-
haired flanks. The predatory eye is riveted.

The broad valley in which the muskoxen graze has the color
and line of a valley in Tibet. I raise my field glasses to draw it
nearer. Beyond the resolution of the ground glass the animals
look darker, the tans of the hills more deeply pigmented, and the
sky at the end of the distant valley is a denser blue. The light
shimmers on them. I recall the observation of a Canadian muskox
biologist: 'They are so crisp in the landscape. They stand out like
no other animal, against the whites of winter or the colors of the
summer tundra.'

I put the glasses back in my lap. A timeless afternoon. Off to
my left, in that vast bowl of stillness that contains the meander-
ing river, tens of square miles of tundra browns and sedge
meadow greens seem to snap before me, as immediate as the
pages of my notebook, because of unscattered light in the dust-
less air. The land seems guileless. Creatures down there take a
few steps, then pause and gaze about. Two sandhill cranes stand
still by the river. Three Peary caribou, slightly built and the silver color of the moon, browse a cutbank in that restive way of
deer. Tundra melt ponds, their bright dark blue waters oblique
to the sun, stand out boldly in the plain. In the center of the
large ponds, beneath the surface of the water, gleam cores of
aquamarine ice, like the constricted heart of winter.

On the far side of the Thomsen other herds of muskoxen graze
below a range of hills in clusters of three or four. In groups of
ten and twelve. I sketch the arrangements in my notebook. Most
remarkable to me, and clear even at a distance of two or three

miles because of the contrast between their spirited, bucking gambols and the placid ambling of the others, is the number of calves. Among forty-nine animals, I count twelve calves. The mind doesn't easily register the sustenance of the sedge meadows, not against the broad testimony of the barren hills and eroded plateaus. It balks at the evidence of fecundity, and romping calves. The muskoxen on the far side of the river graze, nevertheless, on sweet coltsfoot, on mountain sorrel, lousewort, and pendant grass, on water sedge. The sun gleams on them. On the melt ponds. The indifferent sky towers. There is something of the original creation here.

I bring my glasses up to study again the muskoxen in the far valley to the south. Among fifty or sixty animals are ten or fifteen calves. I regard them for a while, until I hear the clattering alarm call of a sandhill crane. To the southwest an arctic hare rises up immediately, smartly alert. To the southeast a snowy owl sitting on a tussock, as conspicuous in its whiteness as the hare, pivots its head far around, right then left. The hare, as intent as if someone had whistled, has found me and fixed me with his stare. In that moment I feel the earth bent like a bow and sense the volume of space between us, as though the hare, the owl, and I stood on a dry lakebed. The moment lasts until the hare drops, becomes absorbed again in the leaves of a willow. The owl returns its gaze to the river valley below.

The indictment of the sandhill cranes continues; I shift my perch so they cannot see me, and their calls cease.

Behind me to the north my four companions, dressed in patterns of color unmistakably human, are at work on a hillside: archaeologists, meticulously mapping the placement of debris at a nineteenth-century Copper Eskimo campsite called PjRa-18, or, informally, the Kuptana site. Like others in the region, this campsite sits near the top of a windswept hill and looks down on well-developed sedge meadows, exceptional muskox pasturage. Scattered over 20,000 square yards at the campsite are more than 27,000 pieces of bone, representing the skeletal debris of about 250 muskoxen. The archaeologists call it a 'death assemblage.' There are also rings of river stones there, which once anchored caribou-skin tents against wind and rain; and the remnants of meat caches built of shale and ironstone slabs, and of charcoal fires of willow twigs and arctic white heather. The lower jaw of an arctic char, eaten a hundred years ago, still glistens with fish oil. A sense of timelessness is encouraged by this primordial evidence of human hunger.

The muskoxen grazing so placidly in the hills to the south and in the sedge meadows to the east, so resplendent with life, are perhaps descendants of these, whose white, dark-horned skulls now lie awry on the land.

BARRY LOPEZ, *Arctic Dreams*, 1986

 *New Naturalists*

*B*Y *the beginning of the twentieth century, the Darwinian scientific revolution had all but swamped traditional natural history. Fieldwork became a diversion, eclipsed by the indoor rigours of laboratory experiment. The discursive essays of writers like Jefferies and Hudson were replaced by the dry, specialized, and often know-all tones of the academic paper. But after the Great War, with its revelations of the fragility of both nature and human values, a new kind of naturalist (or perhaps an old kind in a new guise) began to emerge. They were self-effacing, diligent, and devoted. They moved back to the field and often to much-loved creatures and places, but took with them a new discipline in their approach to observation and deduction. Above all they were fascinated by* behaviour, *and by the interactions of wild creatures with their habitats. It was an atmosphere in which the dedicated amateur could play a role again, and the gap between the worlds of natural history and literature began to close. Many contemporary poets, such as Geoffrey Grigson and Andrew Young, were accomplished naturalists, and relationships with nature were seen as an intrinsic and respectable part of fiction and autobiography, as in the writings of John Moore, E. B. White, and Jocelyn Brooke.*

But the industrial and agricultural upheavals of the 1960s opened new fault-lines. Faced with what seemed to be a critically damaged ecosystem, many writers became polemicists. Others turned to the city as a more potent source of metaphor, or sought refuge in a nostalgic rural past. In Europe at least, only a very few descriptive writers, such as Jane Goodall and J. A. Baker, seemed able to continue celebrating nature's 'minute particulars' and yet convey, just under the surface, a real sense of its vulnerability.

Wildlife biology as a profession interests me. Like the law, my father's vocation, it's one I follow. It's a stepchild among the sciences, however—badly paid, not quite respected, still rather scattered in its thrust and mediocre in its standards, and still accessible to the layman, as the most fundamental, fascinating breakthroughs alternate with confirmation of what has always been common knowledge—akin to that stage of medical research that told us that cigarettes were, yes, 'coffin nails,' and that frying foods in fat was bad for you. Partly because of its romantic bias, as a science wildlife biology has a tragic twist, since the beasts that have attracted the most attention so far are not the possums and armadillos that are thriving but the same ones whose heads hunters like to post on the wall: gaudy giraffes and gorillas, or mermaid-manatees, or 'same-size predators,' in the phrase ethologists use to explain why a grizzly bear regards a man thornily. It's not that the researchers have hurried to study the animals which are disappearing in order to glean what they can, but that the passion that activates the research in the first place is the passion which has helped hound these creatures off the face of the earth. Such men are hunters *manqué.*

Game wardens are also that way, but have the fun of stalking, ambushing, and capturing poachers, while so often the biologist sees his snow leopards, his orangutans, his wild swans and cranes, vanishing through change of habitat right while his study progresses—wondering whether his findings, like other last findings, invulnerable to correction though they soon will be, are all that accurate. Anthropology can be as sad a science when limited to living evidence and a primitive tribe, but the difference is that woodcraft itself is guttering out as a gift, and apart from the rarity now of observers who can get close to a wilderness animal which has not already been hemmed into a reserve, there is the painful mismatch of skills involved in first actually obtaining the data and then communicating it. Scientific writing need only be telegraphic to reach a professional audience, but again and again one runs into experts who have terrible difficulties in setting down even a small proportion of what they have learned. Eagerly, yet with chagrin and suspicion of anybody with the power to do the one thing they wish they could do (suspicion of city folk is also a factor), they welcome television and magazine reporters and interrogators like me, sometimes in order to see their own stories told, but sometimes to try to help save the animals dear to them—as if our weak words might really succeed.

But we observers have a piece missing too; maybe we put on our hiking boots looking for it. Like some of the wildlife experts—or like Lady Chatterley's gamekeeper, who was in retreat when he went into the woods—we don't entirely know why we are there. Not that an infatuation with wild beasts and wild places does us any harm or excludes the more conventional passions like religion or love, but if I were to drive by a thicket of palmettos and chicken trees way down South and you told me that a drove of wild razorback hogs lived back in there, I'd want to stop, get out, walk about, and whether or not the place was scenic, I'd carry the memory with me all day. It's said of a wilderness or an animal buff that he 'likes animals better than people,' but this is seldom true. Like certain pet owners, some do press their beasties to themselves as compresses to stanch a wound, but others are rosy, sturdy individuals. More bothersome to me is the canard that *when I was a man I put away childish things*, and I can be thrown into a tizzy if a friend begins teasing me along these lines. (A sportswriter I know has gone so far as to consult a psychiatrist to find out why a grown man like him is still so consumed by baseball.) Rooting around on river-banks and mountain slopes, we may be looking for that missing piece, or love, religion and the rest of it—whatever is missing in us—just as so often we are doing in the digging and rooting of sex. Anyway, failure as a subject seems more germane than success at the moment, when failure is piled atop failure nearly everywhere, and the study of wildlife is saturated with failure, both our own and that of the creatures themselves.

EDWARD HOAGLAND, 'Bears, Bears, Bears', 1973: from *Heart's Desire*, 1990

April 25th.—To bed in my clothes—the right way for a right field naturalist—and, rising about half-past two, am on the spot an hour afterwards. Some birds were there already. I could make out their darker figures amidst the surrounding darkness, and, as it grew imperceptibly lighter, these began to move swiftly about over the surface of the ground, looking like enlarged rats in the first dim twilight of early morning. Then suddenly they would disappear, changing into pieces of cut turf lying here and there about the land, amidst a number of which the assembly-ground is situated. From this I knew that, even at this early hour, the birds were making excited runs and rushes at one another, between the intervals of which spasmodic energy they lay crouched and motionless on the ground, in a sort of sexual ecstasy. As the morning advanced, the rushes of the birds became swifter and more and more violent, whilst each spring which they made at each other was accompanied with a loud whirring of the wings. Evidently they were actuated by a much more intense and stirring spirit than I had yet seen, and this grew and grew as they were joined by others, and yet others, till the party consisted of a score or more, exclusive of the reeves. Now too it becomes more clearly evident that the whole end and object of the gathering is courtship on the part of the ruffs, choice and acceptance on that of the reeves, that the gathering-ground is essentially the temple of Venus, not of Mars, and that such fighting as there is—violent indeed while it lasts, but never lasting long—is merely incidental to the one all-important purpose of the perpetuation of the species. And this—again it seems clear to me—is entirely dependent on the will of the reeve. As yet, in despite of rufflings, crouchings, prostrations and every form of voluptuous solicitation within the power of the male bird—in spite, too, of battles royal which, though never more than beginning, are yet, by their number and frequency, never ending—nothing has been done in this way.

There are now two reeves that I can see—there may be more in the confusion—and whilst the sun brightens and the frosted grass shines fair and fresh in the morning air, the first union takes place between one of these and the brown bird so prominent in yesterday's proceedings. Then, amidst the feather-waves of this little turbulent sea of desire, I make out five or six reeves, and shortly afterwards there are two more unions, not with the brown one, this time, but with that very other one—he of the blue-black mane and white, flowing head-plumes—which I saw chosen yesterday by a certain reeve who came up to him and touched him on the head with her bill. Thus the two ruffs which have been the first to procure mates have certainly been singled out, and that more than once, by one or more reeves, and so far as I have been able to see—and I think had it been so, I must have seen—they certainly have not owed their good fortune either to fighting or superior 'vigour' of any kind. They are, however—as to this there could hardly be two opinions—two of the finest and handsomest birds on the ground.

The scene now is most interesting, but quite beyond description. Birds dart like lightning over the ground, turn, crouch, dart

again, ruffle about each demure-looking, unperturbed little attraction, spring at each other, and then, as if earth were inadequate
as a medium of emotional expression, rise into the air and dart
round overhead on the wing. The air resounds with the frequent
dull shocks of bodies and the violent whirring of wings. It is all
motion, all energy at the very fever-point of excitement, and then,
all at once, a sudden cessation, almost a sudden death—only the
feathers of each bird's back to be seen, or the tops of their head-
gear or ruff or tail-feathers waving here and there in the wind,
as they lie in tense rigid immobility, like so many little bows of
Ulysses, bent by themselves and ready, each moment, to spring
back. A wonderful drama, truly, of bird-life thus unfolding itself
before me in the early, bright, but bitterly cold morning, whilst
learned ornithologists, all the world over, lie sleeping in their
pleasant beds. But they'll come down all the fresher, to breakfast, and issue Papal bulls against sexual selection, from their
nice, warm, comfortable cathedrals. Thus has truth to wait—
truth, that 'dog that must to kennel,' that 'must be whipped out
when Lady, the brach, may stand by the fire and stink.' In the
midst of it all, but after a considerable interval—for it is long
before things become quieter—there are four more pairings, in
quick succession, between the same brown ruff and a reeve which,
as I cannot identify her, I will not assume to be the same who
chose him yesterday. Possibly, too, there is another nuptial con-
summation in which the blue-maned bird, often before mentioned, plays his part. Of this however I cannot be sure. It is a
business, indeed, when all moves and changes so quickly, to make
out all that goes on. Then, whilst things are at their height, the
whole flock take wing and dart away, to return again in but a
few seconds. These sudden intermittent flights off of all the birds,
together, seem often to be quite instantaneous, and are not reasonably attributable to any extraneous cause. Always untenable,
the leadership theory is here, in the very swirl and vortex of the
sexual maelstrom, preposterous, and the only alternative one, to
that of some kind of subliminal interaction, would seem to be
that each bird reaches just that pitch of excitement which ren-
ders the relief of flight necessary, at one and the same moment
of time.

EDMUND SELOUS, *Realities of Bird Life*, 1927: observations
of a colony of ruffs on the Dutch coast

On a sunny day in the summer of 1929, I was walking rather aimlessly over the sands, brooding and a little worried. I had just done my finals, had got a part-time job, and was hoping to start on research for a doctor's thesis. I wanted very much to work on some problem of animal behaviour and had for that reason rejected some suggestions of my well-meaning supervisor. But rejecting sound advice and taking one's own decisions are two very different things, and so far I had been unable to make up my mind.

While walking about, my eye was caught by a bright orange–yellow wasp the size of the ordinary jam-loving *Vespa*. It was busying itself in a strange way on the bare sand. With brisk, jerky movements it was walking slowly backwards, kicking the sand behind it as it proceeded. The sand flew away with every jerk. I was sure that this was a digger wasp. The only kind of that size I knew was *Bembex*, the large fly-killer. But this was no *Bembex*. I stopped to watch it, and soon saw that it was shovelling sand out of a burrow. After ten minutes of this, it turned round, and now, facing away from the entrance, began to rake loose sand over it. In a minute the entrance was completely covered. Then the wasp flew up, circled a few times round the spot, describing wider and wider loops in the air, and finally flew off. Knowing something of the ways of digger wasps, I expected it to return with a prey within a reasonable time, and decided to wait.

Sitting down on the sand, I looked round and saw that I had blundered into what seemed to be a veritable wasp town. Within ten metres I saw more than twenty wasps occupied at their burrows. Each burrow had a patch of yellow sand round it the size of a hand and, to judge from the number of these sand patches, there must have been hundreds of burrows.

I did not have to wait long before I saw a wasp coming home. It descended slowly from the sky, alighting after the manner of a helicopter on a sand patch. Then I saw that it was carrying a load, a dark object about its own size. Without losing hold of it, the wasp made a few raking movements with its front legs, the entrance became visible and, dragging its load after it, the wasp slipped into the hole.

At the next opportunity I robbed a wasp of its prey, by scaring it on its arrival, so that it dropped its burden. Then I saw that the prey was a honey bee.

I watched these wasps at work all through the afternoon, and soon became absorbed in finding out exactly what was happening in this busy insect town. It seemed that the wasps were spending part of their time working at their burrows. Judging from the amount of sand excavated, these must have been quite deep. Now and then a wasp would fly out and, after half an hour or longer, return with a load, which was then dragged in. Every time I examined the prey, it was a honey bee. No doubt they captured all these bees on the heath, for all the to-and-fro traffic was in the direction of the south-east, where I knew the nearest heath to be. A rough calculation showed that something was going on here that would not please the owners of the bee-hives on the heath; on a sunny day like this several thousand bees fell victims to this large colony of killers!

As I was watching the wasps, I began to realize that here was a wonderful opportunity for doing exactly the kind of field work I would like to do. Here were many hundreds of digger wasps—exactly which species I did not know yet, but that would not be difficult to find out. I had little doubt that each wasp was returning regularly to its own burrow, which showed that they must have excellent powers of homing. How did they manage to find their way back to their own burrow? . . .

PHILANTHUS: SOME OBSERVATIONS

Settling down to work, I started spending the wasps' working day, which lasted from about 8 a.m. till 6 p.m., on the 'Philanthus plains'. We called this part of the sands by that name as soon as we had found out that *Philanthus triangulum Fabr.* was the official name of this bee-killing digger wasp. Its vernacular name was Bee-Wolf.

An old chair, field glasses, notebooks, and food and water for the day were my equipment. The local conditions on the open sands were quite amazing, considering that ours is a temperate climate. Surface temperatures of $43°C$ were not rare and, judging from the response of my skin, which developed a dark tan, I got my share of ultraviolet radiation.

My first job was to find out whether each wasp was really limited to one burrow, as I suspected from the unhesitating way in which the home-coming wasps alighted on the sand patches in front of the burrows. I installed myself in a densely populated quarter of the colony, five metres or so from a group of about twenty-five nests. Each burrow was marked and mapped. Whenever I saw a wasp at work at a burrow, I caught it and, after a short unequal struggle, adorned its back with one or two coloured dots (using quick-drying enamel paint) and released it. These wasps soon returned to work, and after a few hours I had ten wasps, each marked with a different combination of colours, working right in front of me. It was remarkable how this simple trick of marking my wasps changed my whole attitude to them. From members of the species *Philanthus triangulum*, they were transformed into personal acquaintances, whose lives from that very moment became affairs of the most personal interest and concern to me.

NIKO TINBERGEN, *Curious Naturalists*, rev. edn. 1974

It is time that we passed to some of the advantages of size. One of the most obvious is that it enables one to keep warm. All warm-blooded animals at rest lose the same amount of heat from a unit area of skin, for which purpose they need a food-supply proportional to their surface and not to their weight. Five thousand mice weigh as much as a man. Their combined surface and food or oxygen consumption are about seventeen times a man's. In fact a mouse eats about one quarter its own weight of food every day, which is mainly used in keeping it warm. For the same reason small animals cannot live in cold countries. In the arctic regions there are no reptiles or amphibians, and no small mammals. The smallest mammal in Spitzbergen is the fox. The small birds fly away in the winter, while the insects die, though their eggs can survive six months or more of frost. The most successful mammals are bears, seals, and walruses.

Similarly, the eye is a rather inefficient organ until it reaches a large size. The back of the human eye on which an image of the outside world is thrown, and which corresponds to the film of a camera, is composed of a mosaic of 'rods and cones' whose diameter is little more than a length of an average light wave. Each eye has about half a million, and for two objects to be distinguishable their images must fall on separate rods or cones. It is obvious that with fewer but larger rods and cones we should see less distinctly. If they were twice as broad two points would have to be twice as far apart before we could distinguish them at a given distance. But if their size were diminished and their number increased we should see no better. For it is impossible to form a definite image smaller than a wave-length of light. Hence a mouse's eye is not a small-scale model of a human eye. Its rods and cones are not much smaller than ours, and therefore there are far fewer of them. A mouse could not distinguish one human face from another six feet away. In order that they should be of any use at all the eyes of small animals have to be much larger in proportion to their bodies than our own. Large animals on the other hand only require relatively small eyes, and those of the whale and elephant are little larger than our own.

For rather more recondite reasons the same general principle holds true of the brain. If we compare the brain-weights of a set of very similar animals such as the cat, cheetah, leopard, and tiger, we find that as we quadruple the body-weight the brain-weight is only doubled. The larger animal with proportionately larger bones can economize on brain, eyes, and certain other organs.

Such are a very few of the considerations which show that for every type of animal there is an optimum size. Yet although Galileo demonstrated the contrary more than three hundred years ago, people still believe that if a flea were as large as a man it could jump a thousand feet into the air. As a matter of fact the height to which an animal can jump is more nearly independent of its size than proportional to it. A flea can jump about two feet, a man about five. To jump a given height, if we neglect the resistance of the air, requires an expenditure of energy proportional to the jumper's weight. But if the jumping muscles form a constant fraction of the animal's body, the energy developed per ounce of muscle is independent of the size, provided it can be developed quickly enough in the small animal. As a matter of fact an insect's muscles, although they can contract more quickly than our own, appear to be less efficient; as otherwise a flea or grasshopper could rise six feet into the air.

And just as there is a best size for every animal, so the same is true for every human institution. In the Greek type of democracy all the citizens could listen to a series of orators and vote directly on questions of legislation. Hence their philosophers held that a small city was the largest possible democratic state. The English invention of representative government made a democratic nation possible, and the possibility was first realized in the

United States, and later elsewhere. With the development of broadcasting it has once more become possible for every citizen to listen to the political views of representative orators, and the future may perhaps see the return of the national state to the Greek form of democracy. Even the referendum has been made possible only by the institution of daily newspapers.

To the biologist the problem of socialism appears largely as a problem of size. The extreme socialists desire to run every nation as a single business concern. I do not suppose that Henry Ford would find much difficulty in running Andorra or Luxembourg on a socialistic basis. He has already more men on his pay-roll than their population. It is conceivable that a syndicate of Fords, if we could find them, would make Belgium Ltd or Denmark Inc. pay their way. But while nationalization of certain industries is an obvious possibility in the largest of states, I find it no easier to picture a completely socialized British Empire or United States than an elephant turning somersaults or a hippopotamus jumping a hedge.

> J. B. S. HALDANE, 'On Being the Right Size', *Possible Worlds*, 1965

The lizard, still a little faint, allowed me to inspect it and discover two pink and penetrating wounds, the colour of balas rubies, at the base of its tail. Apart from those, it was simply suffering from nervous shock. I decided to give it board and lodging in my clinic cage while it convalesced. Its plump belly and the lack of turquoise-blue markings on its temple informed me that I had been wrong in thinking of it as a he. The *verdelle* took its place in the clinic cage as successor to the tortoise-dying-of-thirst, a tortoise from the Provençal woods, yellow and black, which the long drought had brought, empty, weighing almost nothing, scarcely alive, to my inexhaustible well and my generous watering cans. Before giving him my kitchen garden and its daily sprinkling of dew, its green salads and its slugs, I had given him a room to live in and prescribed a course of overfeeding. To each day, its different needs.

My lizard, once in its cage, was given fresh water, a leafy branch, some dry sand, a few drops of milk, a strip of woollen cloth folded in two for it to sleep and meditate upon, and then I went off in quest of some meal worms. A miracle! I found some. At the mere sight of them, the lizard resumed its grand, dragonish, dress uniform. Swollen, embossed, drawn up on to its stilted front toes, every muscle taut beneath its breastplate of chrysoprase, enamelwork, and melted opal, it arched its back into a hoop and brought its nose down at the whitish worm with all the power of a striking snake. Though it was already caught, already disembowelled, it killed its worm and then rekilled it— just as my French bulldog bitch does with a slipper—and I could hear its horny nose, tock-tock, knocking against the cage floor like a finger in a thimble.

She ate well, and her pink wounds soon healed. She also learned to recognize me, and I was simply flattered by this, rather than surprised. She never hesitated to climb up on to my out-stretched hand, and in the hot noons that boil the blood of lizards lying on their dry flints, she played with me at being a very fierce male lizard. Haughtily perched on the tips of her delicate toes, her belly like a span of a bridge and her scarlet throat full of mock menaces, she took aim at my nose or my cheek and nipped my fingers hard enough to leave her mark on me. From there, we progressed to lunching together. She would lie in all her green and golden brilliance on my tablecloth as the sun stood at its zenith and counselled her to immobility. By this means she re-assured the flies, and then snapped one up from time to time in an electric arabesque. I discovered a passion in her for fresh cream, certain raw fruits, and the syrup of my compotes. She ate too much, smacking her horny lips like a badly brought up child, then lay prostrate and suffering from her gourmandizing. We never know how to set a limit to the favours we receive, we coun-try hicks.

I was afraid of seeing her turn purple and die one day, as a tree frog had once done, before my very eyes, after swallowing an enormous grey fly covered in bristles. I was afraid above all of weakening in the resolution I had made, the best, that no ani-mal should ever in my home forfeit its liberty, its normal chances of life and death, and even that trembling that seizes an untamed creature when it feels that we are watching it.

So I took my plump little lizard, perched on my index finger and confronting the universe around her with an air of defiance, back to the vines. 'If she doesn't run away when I release her', I thought, 'that will mean that I must keep her and take her back to Paris, where she will lie and meditate on the warm stones of the hearth and gaze with reverence at the basket of hot coals. She will lie on the roof of the phonograph and listen to the music vibrating in her belly; she will run vertically up the length of the curtains, and keep me company clinging to the warmth of my lampshade . . .'

As I was thinking these thoughts, I laid her down in the hollow of a tiny track among the vines. The same sand, in that spot, nourishes the vines and forms a filter for the sea. Cold in the dawn, at noon it will burn the sole of an unshod foot. It sends the lizard into ecstasy, and the tortoise, when the sun is at its height, cannot contain itself. Because I have always remained naïve in my relationships with the animal world, I thought at first that the lizard had already made its choice, between me and . . . all the rest. For she remained quite calm, her tail held in a resplendent curve against her flank. I scratched her head and her throat swelled up in friendship, then I stopped scratching her head, so as to leave her free to make her choice. And then . . . And then nothing more happened. The spot where the lizard lay glittering was suddenly void. I had, it is true, perceived a green flash, the beginning of a tremendous feat of speed. But our senses are slow, and the fine, floury, shifting sand had retained no imprint. But at least that lizard carries with it, symmetrically positioned on each side of the plump tail, and perhaps suspect to all other members of that tribe arrayed in quartzy green, opal, and gold, the traces of a human solicitude that supposed itself to be disinterested.

COLETTE, 'A Lizard', *Prisons et Paradis*, 1932; tr. Derek Coltman, 1966

The finest blackberries I have ever seen grow upon a heath about two miles distant from Peverel. A much-frequented main road from London to the sea cuts through it, and motorists, halting for a picnic lunch, marvel at the exceptional size of the blackberries there. It is no wonder, really, for those particular brambles have had a costly pruning: it took a European war to bring them to their present perfection.

For five years this heath was the site of a camp. Where from time immemorial had been nothing but briar and bracken, row upon row of wooden huts, churches, shops, and theatres sprang up in a week or two. Where only the lapwing had cried or the skylark sung, the drill-sergeant's word of command rang out. The whole place became a populous town.

Tens of thousands of Canadian soldiers sojourned there. One contingent after another arrived, the men often soaked with rain or moiled with heat, and always cramped from the close quarters of war-time transport. There they had a breathing space, saw a little of the 'old country', and learned to love it. They were drilled upon those open spaces so flattened by their feet that even now the heather has scarcely begun to grow again. Then each battalion in its turn passed singing along that same main road to its fate.

In the course of these operations such flower and bramble roots as were left were cut back to earth and received a dressing of all kinds of camp residue. When auction sales and motor lorries had removed the last vestiges of the buildings, and a small army of workmen had laboured for months removing rubbish and filling in holes, Nature set to work to heal the scars; and almost the first growth was the long green shoots of the blackberry brambles. The fruit, when it followed, was of the finest—cultivated fruit indeed—and cultivated at what tremendous cost!

Now the bushes are of full size again. Bracken has grown up and filled the rents made by bomb-practice. The heather has returned in waves, a purple sea. Very soon all will be as it had been for countless ages before war broke out, and only the avenue of maple trees the Canadians planted by the roadside will mark any difference between that heath and a score of others by that same roadside . . .

> FLORA THOMPSON, in *Catholic Fireside* (c.1925): from *The Peverel Papers*, ed. Julian Shuckburgh, 1986. Between 1922 and 1927 Flora Thompson, author of *Lark Rise to Candleford*, wrote a regular column on nature and country life for the weekly magazine *Catholic Fireside*

A word I find sinister in a different fashion is the naturalists' term for a caterpillar: larva, which comes from Latin, in which language it has an uncomfortable, spine-chilling, prickling-at-the-back-of-the-neck connotation. It means the walking spirit of a dead person, but carries the implication that the unresting dead one is in pursuit of the living. Because such a spirit is faceless, the word in Latin acquired the additional meaning of 'a mask.'

By a most daring fancy, our old naturalists adopted it as the scientific name for a caterpillar; because such a creature wears a disguise, the future insect is not recognisable in the present grub, its form is a 'mask' which will one day be cast off. The name dates, of course, from the days before science and the humanities set themselves at odds; a good natural historian was generally a fair classical scholar, and he used the classics to make his communications concerning science more vivid, imaginative, logical and accurate. So science and poetry coexisted—the use of 'larva' for a caterpillar, and of 'pupa' for the next metamorphosis, is a truly poetic employment of words. The great naturalist Linnaeus was the first to use 'pupa' in this sense, in 1758. It is simply the Latin word for girl-child, hence for a doll; and Linnaeus' adoption of it was a stroke of genius, as you will realise if you look at the underside of a moth's pupa and see the shape of its face, eyes and embryonic wings like little arms sedately crossed in front of its body, all wrapped as if in swaddling-clothes which emphasise its likeness to a doll. Other words from the Latin *pupa* or French *poupée* are puppy (as it were a toy dog, a dog-doll) and pupil, in both senses—that of the eye is so called because of the tiny image reflected there. From *poupette*, a baby doll, we get puppet and also poppet, the term of endearment.

JOHN MOORE, You English Words, 1961

June took me away and the house was left empty. And when I returned in July, I knew at once that something was wrong. Martins, by the dozen, flickered over the roof and ransacked the evening sky for food; but they were not my martins; or, at least, they had nothing to do with the nest over my door. At first I thought perhaps the birds had hatched their eggs and the fledglings had already flown; but when I looked at other nests, in near-by eaves, where the martins had been busily building long before mine came on the scene, I was convinced that this could not be so—for there the youngsters were still thrusting their yellow gapes out of the nest-hole, clamorous for food and yet more food.

It was a straw sticking out of my martin's nest that first betrayed the truth. And then a sparrow-face appeared at the opening, chirruped saucily and disappeared inside again.

I know there are people who have a special affection for sparrows: they are so impudently indomitable, thriving where more delicate birds would perish, insouciant, undeniable, the *gamin* of the skies. Well, I give them all that and more; but still I do not like them. And if I did not like them before, how shall I like them now, when they have turned my martins out of house and home? My thatch had been wired to keep them away: but no, that did not defeat them. Those sparrows were determined to be my guests. If they could not get into the thatch, at least they could turn the martins out of their nest and make themselves at home there. Anyway, who *are* these fanciful weaklings that must needs hurry off south directly the cold weather comes? Come along, we'll show them who's master here: the nasty foreigners!

And so—well, the long and the short of it was that that cheeky face peeping out of my martins' nest made me see red. Saint Francis, who loved the house-sparrow equally with the nightingale, forgive me: I took a pole and poked the nest, till it fell in broken shards to the ground. That will teach you, I said; though of course it didn't.

Next morning the impossible seemed to have happened. My martins were returning! The same swooping up under the eaves, the same soft chuckling as of a hidden burn, the same giddy spiral-flights past my window. Then they would cling against the plaster, probing with their needle-beaks the ruins of their home. The sun shone on the steel-blue of their backs, and breast and throat gleamed like snow. Their long, tapering wings were crossed above them, like the slim, folded wings of Fra Angelico's angels. And they chattered unceasingly. Out of the blue sky more martins appeared and joined them, pricking the morning with a multitude of small cries, glissading down the air and briefly inspecting the ruins as they passed. It seemed as if my martins had called all their friends and neighbours to come and help them to decide the momentous question whether, even at this late stage, they should build a second house and rear a second family. What finally decided them, I do not know; but after that one morning of frenzied consultation, they never came near the place again.

Next year, the same thing happened. With the celandines came the martins. A few feet away from last year's sad reminder, they built another nest. Again there was the bright eye looking out at me every time I passed. Again I was called away. And again the ragamuffin sparrows turned them out. So now I am wondering will the martins give me a third chance to get rich? Will they try again next spring?

Next spring? But nobody dares to think of next spring and what the war may bring for us with the warmer weather and the lengthening light. Thanks be, then, for the respite of these shivering days that shut away the fear of what may be.

> C. HENRY WARREN, *England Is a Village*, 1940: a classic portrait of an Essex village in wartime

Some people might consider an apology necessary for the appearance of a book about birds at a time when Britain is fighting for its own and many other lives. I make no such apology. Birds are part of the heritage we are fighting for. After this war ordinary people are going to have a better time than they have had; they are going to get about more; they will have time to rest from their tremendous tasks; many will get the opportunity, hitherto sought in vain, of watching wild creatures and making discoveries about them. It is for these men and women, and not for the privileged few to whom ornithology has been an indulgence, that I have written this little book.

I shall be very pleased if anybody who reads it becomes interested in birds.

> JAMES FISHER, Preface to *Watching Birds*, 1940

Beginning a decade ago, we have trapped and banded most of the chickadees on our farm each winter. In early winter, the traps yield mostly unbanded birds; these presumably are mostly the young of the year, which, once banded, can thereafter be 'dated.' As the winter wears on, unbanded birds cease to appear in the trap; we then know that the local population consists largely of marked birds. We can tell from the band numbers how many birds are present, and how many of these are survivors from each previous year of banding.

65290 was one of 7 chickadees constituting the 'class of 1937.' When he first entered our trap, he showed no visible evidence of genius. Like his classmates, his valor for suet was greater than his discretion. Like his classmates, he bit my finger while being taken out of the trap. When banded and released he fluttered up to a limb, pecked his new aluminum anklet in mild annoyance, shook his mussed feathers, cursed gently, and hurried away to catch up with the gang. It is doubtful whether he drew any philosophical deductions from his experience (such as 'all is not ants' eggs that glitters'), for he was caught again three times that same winter.

By the second winter our recaptures showed that the class of 7 had shrunk to 3, and by the third winter to 2. By the fifth winter 65290 was the sole survivor of his generation. Signs of genius were still lacking, but of his extraordinary capacity for living, there was now historical proof.

During his sixth winter 65290 failed to reappear, and the verdict of 'missing in action' is now confirmed by his absence during four subsequent trappings.

At that, of 97 chicks banded during the decade, 65290 was the only one contriving to survive for five winters. Three reached 4 years, 7 reached 3 years, 19 reached 2 years, and 67 disappeared after their first winter. Hence if I were selling insurance to chicks, I could compute the premium with assurance. But this would raise the problem: in what currency would I pay the widows? I suppose in ants' eggs.

I know so little about birds that I can only speculate on why 65290 survived his fellows. Was he more clever in dodging his enemies? What enemies? A chickadee is almost too small to have any. That whimsical fellow called Evolution, having enlarged the dinosaur until he tripped over his own toes, tried shrinking the chickadee until he was just too big to be snapped up by flycatchers as an insect, and just too little to be pursued by hawks and owls as meat. Then he regarded his handiwork and laughed. Everyone laughs at so small a bundle of large enthusiasms.

The sparrow hawk, the screech owl, the shrike, and especially the midget saw-whet owl might find it worth while to kill a chickadee, but I've only once found evidence of actual murder: a screech-owl pellet contained one of my bands. Perhaps these small bandits have a fellow-feeling for midgets.

It seems likely that weather is the only killer so devoid of both humor and dimension as to kill a chickadee. I suspect that in the chickadee Sunday School two mortal sins are taught: thou shalt not venture into windy places in winter, thou shalt not get wet before a blizzard.

I learned the second commandment one drizzly winter dusk while watching a band of chicks going to roost in my woods. The drizzle came out of the south, but I could tell it would turn northwest and bitter cold before morning. The chicks went to bed in a dead oak, the bark of which had peeled and warped into curls, cups, and hollows of various sizes, shapes, and exposures. The bird selecting a roost dry against a south drizzle, but vulnerable to a north one, would surely be frozen by morning. The bird selecting a roost dry from all sides would awaken safe. This, I think, is the kind of wisdom that spells survival in chickdom, and accounts for 65290 and his like.

The chickadee's fear of windy places is easily deduced from his behavior. In winter he ventures away from woods only on calm days, and the distance varies inversely as the breeze. I know several wind-swept woodlots that are chickless all winter, but are freely used at all other seasons. They are wind-swept because cows have browsed out the undergrowth. To the steam-heated banker who mortgages the farmer who needs more cows who need more pasture, wind is a minor nuisance, except perhaps at the Flatiron corner. To the chickadee, winter wind is the boundary of the habitable world. If the chickadee had an office, the maxim over his desk would say: 'Keep calm.'

His behavior at the trap discloses the reason. Turn your trap so that he must enter with even a moderate wind at his tail, and all the king's horses cannot drag him to the bait. Turn it the other way, and your score may be good. Wind from behind blows cold and wet under the feathers, which are his portable roof and air-conditioner. Nuthatches, juncos, tree sparrows, and woodpeckers likewise fear winds from behind, but their heating plants and hence their wind tolerance are larger in the order named. Books on nature seldom mention wind; they are written behind stoves.

I suspect there is a third commandment in chickdom: thou shalt investigate every loud noise. When we start chopping in our woods, the chicks at once appear and stay until the felled tree or riven log has exposed new insect eggs or pupae for their delectation. The discharge of a gun will likewise summon chicks, but with less satisfactory dividends.

What served as their dinner bell before the day of axes, mauls, and guns? Presumably the crash of falling trees. In December 1940, an ice-storm felled an extraordinary number of dead snags and living limbs in our woods. Our chicks scoffed at the trap for a month, being replete with the dividends of the storm.

65290 has long since gone to his reward. I hope that in his new woods, great oaks full of ants' eggs keep falling all day long, with never a wind to ruffle his composure or take the edge off his appetite. And I hope that he still wears my band.

ALDO LEOPOLD, '65290', *A Sand County Almanac*, 1949

The song is not only a hymn to the beloved one. It expresses the complex, multifarious emotions that the male, under the imperious or latent influence of the passion lifting him out of himself, cannot confine in a simple cry: assertion of himself, of his vigour and beauty, of his happiness at being alive and at having his place in nature. Thus, at the slightest suspicion of an unfamiliar presence, the Cetti's Warbler fills the reeds with the uproar of his irritated and blustering song. Keeper of his canton, the Robin lets everyone know, in infinitely varied phrases, that he is in possession of the territory and that he intends to remain master there. For singing, to a bird, is throwing a challenge, too. Before the timid females, rivals affront each other with their voices. From one tree top to another the male Chaffinches ceaselessly toss each other their triumphal refrain, as if they were seeking the last breath of their adversaries. In the depths of the hawthorn bushes two Nightingales, face to face, listen to each other sing in turn as if each were trying to carry off the secret that makes singing more beautiful. Over the fields and woods the males' boasting rises thus, loud or discreet, sweet or harsh, repeated from place to place. At the limits of the species' habitat where the struggle of voices becomes attenuated between isolated rivals, the singing loses its strength and beauty.

Indeed, this beauty is not primitive. Descendants of the Saurians by the winged and lizard-tailed fossil, Archaeopteryx, the birds, having gained dry land after the miry marshes of the secondary epoch, and then having raised themselves into the air, preserve in their voices the traces of the croaking of their ancestors. Even in the finest artists themselves the atavistic taint reappears at times. The Nightingale interrupts his most beautiful stanzas with a 'carr' that one might say had come from the flabby throat of a Batrachian. In the Blackbird, the flaw in the precious metal, a guttural note, is found at the end of his whistled phrase. In the Thrush it slips in among the purest cadences in hard, sharp sounds. The Melodious Warbler starts his song on three raucous notes. The sonorous cascade of the Lark, the intimate sweet song of the Bullfinch, the clear, silvery one of the Linnet, all bear, at moments, the mark of the original tare.

Each year, in the springtime, the bird works to strip his song of the primitive stain. The finer the artist, the harder the labour. For weeks his throat must be made more supple, and each day the sounds issue forth a little purer. In January, on mild evenings, the Blackbird practises before sunset. About the same time the Lark, on short flights, drops a few bits of song. On their return to the nesting country in March and April the Blackcap and the Nightingale are apprentices seeking the accents of the year before. The young male, singing for the first time, must recall the paternal voice to which he listened last summer when he was still crouched in the bottom of the nest. Here individual temperament and virtuosity assert themselves in the effort toward perfection. Under the apparent uniformity of phrases, cadences, and tone of the same species, nothing is more plastic than the singing of a bird.

> JACQUES DELAMAIN, *Why Birds Sing*, n.d.; tr. Ruth and Anna Sarason, n.d. *c*.1950

'The best thing a bird-watcher can do,' we agreed, 'is to keep right away from the bee-eaters.' Sadly, if a little self-righteously, we parted.

But man—and woman—is a contradiction. The readiness for self-sacrifice which marks one hour may be forgotten or ignored next day. Since the last hours of July when three observers had been astonished by the sight of five bee-eaters about this Sussex sand-pit, scores of ornithologists had seen the birds. By the fifth day of August, two nests had been discovered, larger and more egg-shaped than the neighbouring burrows of the sand martins. Could we not ask to be allowed a brief distant glimpse of these most beautiful of birds?

The official watcher of the Royal Society for the Protection of Birds, on guard near the nests, was as courteous and kind as the police had been and just as firm. Yes, we might see the bee-eaters. 'But don't walk beyond those cows. And please sit down and keep quiet.'

We sat down. We sat for the best part of two wonderful hours, amazed at the brilliance of the bee-eaters' plumage, astonished by the varied skill of their flight, startled by the beautiful liquid tones of their voices. Their flute-like cries, reminiscent of the song of waders, made us imagine ourselves at some southern estuary—until one bee-eater began to flutter after flies like a spotted flycatcher, another hovered like a hawk, and a third swept across the fields in the manner of a mistle-thrush. As we watched, one of the two juveniles pounced upon an unsuspecting dragonfly; another caught a bumble bee and, after holding it for a moment or two, casually tossed it into the air and caught it again. Presumably this was the bee-eater's well-known 'de-stinging' operation, the stings of insects being discarded in flight before the prey is eaten.

For a time we watched the bee-eaters hawking for insects over the blue waters of a reed-fringed pool where moorhens and mallard were feeding. Flying ten and twenty feet above the irises and rushes, the birds would suddenly fall upon a dragonfly as it patrolled the boundaries of its territory, or dart away in pursuit of a wandering red admiral or small tortoise-shell butterfly.

The prey were not easily obtained. Many times we heard the loud 'snap' as the meeting of the mandibles betrayed the failure of the hunter to seize his victim. Bee-eaters do not lack persistence, though, and the defeated bird would quickly rush away in pursuit of his prey, banking steeply and slipping groundwards like a swallow at play, or sweeping across the sky with the zest of a young martin.

For almost twenty minutes the birds disappeared from view and we began to wonder if we should ever see them again. Then their faint cries came floating out of the wind, prompting us to glance southwards where the clearcut crest of the Downs, backed by moving mountains of cumulus cloud, seemed to throw back the haunting calls of the birds. Next moment they were sweeping over our heads, descending with insect morsels to the young nestlings lurking within the burrows cut into the face of the cliff . . .

One September morning, as the sunlight sent the autumn haze withdrawing into the shelter of hedgerows and ditches, revealing clusters of gossamer on every spray of thorn and gorse, I learnt that the bee-eaters had flown from their sand-pit and not returned. The instinctive call to migrate had driven them—where?

We do not know, though bee-eaters commonly winter in Africa and Arabia. What is certain is that our Sussex bee-eaters were last seen on September 24, flying southwards into the unknown.

And the observer, by a curious stroke of fate, happened to be E. A. Packington, who was the first person to record them early in the breeding season.

Yet though the birds had flown with the departure of summer, their influence on the people of the neighbourhood still lingers. I can testify to this claim, for every week I visit a county second-ary school in the area and observe the activities of some thirty young naturalists from the villages of the district. Year by year, as I make the acquaintance of a fresh group of junior naturalists, a hand will be raised, an expression of excitement and awe mark a keen young face, and I hear: 'Please, sir, in the village where I live, we had bee-eaters.'

The presence of the birds, if only for a season, has attained legendary status and become woven into the fabric of the local folklore, something as real—and as unreal—to a growing boy as the attempt of Guy Fawkes to blow up Parliament.

In many cases the sight of these exotic birds so impressed itself on six- and seven-year-old minds that schoolboys of twelve and thirteen speak as if they saw them yesterday.

GARTH CHRISTIAN, 'The Bee-Eaters', *Down the Long Wind*, 1963

I often wonder how many amateur naturalists reflect on the inter-esting things they have seen or heard in their wanderings and, for some reason, failed to record at the time—still less write up for an appropriate journal. The moralists are always telling us that 'confession is good for the soul', and it may well be in this context. Recently I have indulged in a period of reflection on my own shortcomings, weeping silent tears over intriguing observa-tions I have made over the years. Some might have been of value, had I not been lazy or forgetful. Amateur naturalists are inclined to be a little superior about their amateurism, referring to their qualified and professional colleagues as hidebound, sceptical and too serious-minded. Even if there is a grain of truth in these stric-tures, we all too frequently forget the value of a disciplined train-ing, part of which concerns the prime importance of noting down all incidents observed.

I was made to welter in self-reproach the other day when a friend who knew I was interested in hedgehog behaviour asked me if I had ever seen one walking in circles, apparently aimlessly, and keeping it up for some time. I replied rather casually that I had, and thought no more about it until, a few days later, I was brought down to earth with a bump. I received a letter from a distinguished scientist who is a student of animal behaviour. He told me that my friend had passed on my remarks about the hedgehogs; and he asked for full details, with any references to this curious phenomenon I had read. I hung my head in shame when I realised that, though I had observed similar hedgehog perambulations on three occasions, I had taken insufficient notice and, worse still, had not written down the dates, times and other details at the earliest possible moment. I was able to give only such information as I could recollect. My memory is good; but that is not enough.

Another case also concerns hedgehogs. For some years Dr Maurice Burton has been investigating the curious self-anointing in which they indulge; and he was discussing it with me, describing his experiments with tame hedgehogs. He said he wished he could get information from someone who had seen this behaviour in the wild. I was at once reminded of the time, some years before, when I had without doubt witnessed it. But I then knew nothing of self-anointing and had taken the antics of the urchin for an ailment of some kind; as I had a dog with me, I was more concerned to keep him away from a possible source of infection than I was with the hedgehog's behaviour. Once again I was able to give only personal testimony without data of value.

Then there was the time I first saw a pile of dead frogs near a spawning-pond, each with its hind legs clearly eaten by some predator. I did make a note of this, but failed to count the corpses and to see how many were males and how many females. Later, in discussions with Dr Malcolm Smith and Frances Pitt, I learned that this was the work of a stoat. Dr Smith wanted to know whether there were signs of eggs which had been left severely alone, and I had not troubled to look. I learned a lesson from this and have since seen a stoat, also a brown rat, at work of this kind; and I have heard of water voles behaving in the same way. Always the remnants of spawn have been quite untouched, probably because it is distasteful or even poisonous.

One of the most intriguing sights, which I also failed to write up fully, occurred when I was studying the marsh frog in the dykes on the Romney Marsh. It was after midnight, and the temperature was so cold that my companions and I were all wearing thick sweaters, mufflers and so on even though it was late May. We were using electric torches to spot the frogs in the water, which was equally cold. Suddenly, in the light-beam, three of us saw a grass snake in the middle of a dyke, its head raised, very alert and clearly on the hunt among the frogs. We caught the snake, but the interest lay in the fact that it was feeding at night and at a temperature which one would have expected to cause all good serpents to lie up in some snug spot on land. To be fair, we did make some notes; but, so far as I know, none of us wrote up the incident for publication in some suitable journal. On this occasion, I am happy to record, only one other amateur was present; all the rest were scientists of high degree, including a professor.

It is a little mollifying to reflect that even the mighty have their lapses. Even so it is good that all field naturalists should remember to record with as much detail and as soon as possible even those incidents which appear insignificant at the time. I wonder how many valuable records have been lost through our sins of omission?

MAXWELL KNIGHT, 'Sins of Omission', *The Countryman* magazine; from *The Countryman Wildlife Book*, 1969

When Mary came to dinner we had not seen her for some years. Shortly after the war she and a friend, both just out of the ATS, had come to work on one of the farms, partly with the cows and partly on a market-gardening enterprise which I had started at about that time. A pretty girl, she had grown into a beautiful woman with just a tinge of grey in the dark hair over a young face. During the early summer I had experimented with two eighteen-year-old blondes, and here was an older brunette to act as the control necessary in any carefully designed scientific investigation—and an animal-loving brunette at that.

It was a warm evening, like all the summer evenings of 1959, and the windows and the door were wide open, with the smell of tobacco plants drifting in. By great good fortune there came in, too, a pipistrelle which was shepherded into the outer hall and caught with a butterfly net. Struggling, squeaking and biting furiously, it was obviously a better subject for experiment than the tame one I had in a cage, so I lifted up a handful of Mary's curly hair and popped the bat in. I had done the same with four different species of bat and the blondes, and the earlier results were repeated: the pipistrelle scrambled up over the top of Mary's head without getting entangled in any way and took flight out of the door, back into the night.

It is, of course, almost axiomatic that women's hair has an irresistible attraction for bats and that, once in contact, the two become so inextricably entangled that they can be separated only with scissors wielded by a man. The stories of such occurrences are much like those of the Indian rope trick: one's informant knows somebody, who knows somebody else, whose first cousin knew a girl . . . It is extraordinary how they grow. I was told by a descendant of a long line of country parsons—country parsons of the time when they and country squires had the education, leisure and inclination to indulge in scientific and literary pursuits—that there had always been a tradition in her family that some time in the middle of last century a bat flew into the hair of her great-aunt Nellie while she was sitting in front of the fire with her feet in a mustard bath. Mustard baths and bats in the house are, of course, almost mutually exclusive: the first is a by-product of November fogs, the second of high summer. Yet this story grew up in a family with a strong tradition of critical scientific observation. When bats are said to fly into the hair, reason flies out of the window.

I must confess to considerable vicarious pride that in 1959, the centenary year of the publication of Darwin's *Origin of Species*, three women from my own county were ready to offer themselves thus as martyrs in the sacred cause of science to test the truth of a hoary superstition. It is much to be hoped that some stout-hearted West Country women will be found to experiment with the greater and lesser horseshoe bats, which are not found in Suffolk; we have only Vespertilionids.

EARL OF CRANBROOK, 'After Dinner Experiment', *The Countryman Wildlife Book*, 1969

I guess I have watched my [rac]coon descend the tree a hundred times; even so, I never miss a performance if I can help it. It has a ritualistic quality, and I know every motion, as a ballet enthusiast knows every motion of his favorite dance. The secret of its enchantment is the way it employs the failing light, so that when the descent begins, the performer is clearly visible and is a part of day, and when, ten or fifteen minutes later, the descent is complete and the coon removes the last paw from the tree and takes the first step away, groundborne, she is almost indecipherable and is a part of the shadows and the night. The going down of the sun and the going down of the coon are interrelated phenomena; a man is lucky indeed who lives where sunset and coonset are visible from the same window.

The descent is prefaced by a thorough scrub-up. The coon sits on her high perch, undisturbed by motorcars passing on the road below, and gives herself a complete going over. This is catlike in its movements. She works at the tail until it is well bushed out and all six rings show to advantage. She washes leg and foot and claw, sometimes grabbing a hind paw with a front paw and pulling it closer. She washes her face the way a cat does, and she rinses and sterilizes her nipples. The whole operation takes from five to fifteen minutes, according to how hungry she is and according to the strength of the light, the state of the world below the tree, and the mood and age of the kittens within the hole. If the kittens are young and quiet, and the world is young and still, she finishes her bath without delay and begins her downward journey. If the kittens are restless, she may return and give them another feeding. If they are well grown and anxious to escape (as they are at this point in June), she hangs around in an agony of indecision. When a small head appears in the opening, she seizes it in her jaws and rams it back inside. Finally, like a mother with no baby-sitter and a firm date at the theater, she takes her leave, regretfully, hesitantly. Sometimes, after she has made it halfway down the tree, if she hears a stirring in the nursery she hustles back up to have another look around.

A coon comes down a tree headfirst for most of the way. When she gets within about six feet of the ground, she reverses herself, allowing her hind end to swing slowly downward. She then finishes the descent tailfirst; when, at last, she comes to earth, it is a hind foot that touches down. It touches down as cautiously as though this were the first contact ever made by a mammal with the flat world. The coon doesn't just let go of the tree and drop to the ground, as a monkey or a boy might. She steps off onto my lawn as though in slow motion—first one hind paw, then the other hind paw, then a second's delay when she stands erect, her two front paws still in place, as though the tree were her partner in the dance. Finally, she goes down on all fours and strides slowly off, her slender front paws reaching ahead of her to the limit, like the hands of an experienced swimmer.

I have often wondered why the coon reverses herself, starting headfirst, ending tailfirst. I believe it is because although it comes naturally to her to descend headfirst, she doesn't want to arrive on the ground in that posture, lest an enemy appear suddenly and catch her at a disadvantage. As it is, she can dodge back up without unwinding herself if a dog or a man should appear.

Because she is a lover of sweet corn, the economic status of my raccoon is precarious. I could shoot her dead with a .22 any time I cared to. She will take my corn in season, and for every ear she eats she will ruin five others, testing them for flavor and ripeness. But in the country a man has to weigh everything against everything else, balance his pleasures and indulgences one against another. I find that I can't shoot this coon, and I continue to plant corn—some for her, what's left for me and mine—surrounding the patch with all sorts of coon baffles. It is an arrangement that works out well enough. I am sure of one thing: I like the taste of corn, but I like the nearness of coon even better, and I cannot recall ever getting the satisfaction from eating an ear of corn that I get from watching a raccoon come down a tree just at the edge of dark.

E. B. WHITE, 'Coon Tree', *New Yorker*, 1956: from *Essays of E. B. White*, 1977

The sun was very low; the shadow of the house lay long and dark across the grass and the rushes, while the hill-side above glowed golden as though seen through orange lenses. The bracken no longer looked green nor the heather purple; all that gave back their own colour to the sun were the scarlet rowan-berries, as vivid as venous blood. When I turned to the sea it was so pale and polished that the figures of the twins thigh-deep in the shallows showed in almost pure silhouette against it, bronze-coloured limbs and torsos edged with yellow light. They were shouting and laughing and dancing and scooping up the water with their hands, and all the time as they moved there shot up from the surface where they broke it a glittering spray of small gold and silver fish, so dense and brilliant as to blur the outline of the childish figures. It was as though the boys were the central décor of a strangely lit Baroque fountain, and when they bent to the surface with cupped hands a new jet of sparks flew upward where their arms submerged, and fell back in brittle, dazzling cascade.

When I reached the water myself it was like wading in silver treacle; our bare legs pushed against the packed mass of little fish as against a solid and reluctantly yielding obstacle. To scoop and to scatter them, to shout and to laugh, were as irresistible as though we were treasure hunters of old who had stumbled upon a fabled emperor's jewel vaults and threw diamonds about us like chaff. We were fish-drunk, fish-crazy, fish-happy in that shining orange bubble of air and water; the twins were about thirteen years old and I was about thirty-eight, but the miracle of the fishes drew from each of us the same response.

We were so absorbed in making the thronged millions of tiny fish into leaping fireworks for our delight that it was not for some minutes that I began to wonder what had driven this titanic shoal of herring fry—or soil, as they are called in this part of the world—into the bay, and why, instead of dispersing outwards to sea, they became moment by moment ever thicker in the shallows. Then I saw that a hundred yards out the surface was ruffled by flurries of mackerel whose darting shoals made a sputter of spray on the smooth swell of the incoming tide. The mackerel had driven the fry headlong before them into the narrow bay and held them there, but now the pursuers too were unable to go back. They were in turn harried from seaward by a school of porpoises who cruised the outermost limit of their shoals, driving them farther and farther towards the shore. Hunter and hunted pushed the herring soil ever inward to the sand, and at length every wavelet broke on the beach with a tumble of silver sprats. I wondered that the porpoises had not long since glutted and gone; then I saw that, like the fry and the mackerel that had pursued them into the bay, the porpoises' return to the open waters of the sound was cut off. Beyond them, black against the blanched sunset water, rose the towering sabre fin of a bull killer whale, the ultimate enemy of sea creatures great and small, the unattackable; his single terrible form controlling by its mere presence the billions of lives between himself and the shore.

GAVIN MAXWELL, *Ring of Bright Water*, 1960

As I begin to descend from the ridge, one of the huge birds flies up, its heavy wingbeats rending the air in sharp, loud swooshes.

Another, another—and then fifteen or more scatter from the same place on the ground among the birches in front of me. They vanish like black ghosts in the fog among the white birch trunks. Silence returns.

The ravens had been feasting on the remains of a moose (one left by a poacher that had been all but totally covered up by brush and logs). The remaining meat is still fresh. I remove the hide, cut off a few chunks of meat, make a small fire up on the ledges by the spruces, and prepare my simple meal. I wait to watch the birds. There is no greater pleasure than eating roasted moose while resting under a spruce and contemplating ravens.

BERND HEINRICH, *Ravens in Winter*, 1990

One year I thought I would go bird-watching; so I bought a pair of field-glasses, and with these slung over my shoulder I used to set off. I hoped my neighbours would think I was an ornithologist; what they thought was that I had taken to attending race-meetings. But as I had done nothing to suggest an interest in plants, I was surprised one day when a stranger called and said, 'I understand you are a botanist'. When I told him he was mistaken, he rose to go, but 'Sit down', I said, and we had a pleasant talk. He told me he was well over seventy, was unable to get about much, and would never again climb a mountain; could I— and here the object of his visit appeared—could I send him any plants for his herbarium? I was so touched by this appeal, that a few weeks later, climbing in the Highlands, I picked the very rare Snow Gentian. Wrapping it carefully in moss, and adding a few common flowers that I thought would remind him of the mountains, I sent it with a letter in which I said I hoped I was giving him a pleasant surprise. There was no doubt about the surprise, for he wrote back with strained politeness, thanking me for the common flowers, but saying nothing of the Snow Gentian. I knew what had happened; he had flung it away unnoticed with the moss. That's a lesson to me, I thought; I shall send no more plants to herbariums. How much lighter would be my conscience to-day, if I had kept to that wise decision!

A herbarium is a place, not where plants grow, but where they are buried, a collection of their dried unlovely corpses. Necessary for a botanist no doubt, to ordinary people it should be a shocking sight, like that Booth Museum of Birds at Brighton that

Hudson, lover of birds, refused to view. But perhaps I am prejudiced. When Pepys visited Evelyn, in the interval between hearing

part of a play or two of his making, very good, but not as he conceits them,

and hearing

some little poems of his own, that were not transcendent,

he was shown a herbarium, of which he says:

He showed me his Hortus Hyemalis; leaves laid up in a book of several plants kept dry, which preserve their colour, however, and look very finely, better than any Herball.

Sometimes wild flowers are collected, not to be put in a hortus siccus, as a herbarium is also curiously called, but to be grown in a garden. Many of them die, of course; they are like sea-birds that cannot be tamed; they have not the patience of the Aspidistra, content to exchange the dimness of an Asian forest for the dinginess of a cheap boarding-house. Even if they do survive, they look out of place. A garden is a kind of garden party, where everyone is dressed for the occasion . . .

One day I went to Loch Lubnaig to look for that rare plant, Least Water-lily. I thought I saw it, but could not be sure that what I saw was not the common Yellow Water-lily. As it grew

In dangerous deeps, yet out of danger's way,

I started to undress. A road runs by the loch-side, and on that summer afternoon cars kept passing up and down; but as I could not see the people in the cars, I imagined they could not see me. 'In any case it's their look-out', I said, perhaps more truthfully than I knew. I waded towards the plant and was standing waist-deep in water when I saw an open charabanc, full of holiday-makers, sailing down the road. I felt that was a different matter, and tearing off a flower I hurried back to the shore. My struggles to pull over wet shoulders a shirt that offered determined resistance were watched with interest from the approaching charabanc, and when just in time I succeeded, the hearty holiday-makers rose from their seats and, so to speak, 'greeted the unseen with a cheer'. Some people, I know, despise my methods of botanizing; yet few botanists have had their efforts publicly applauded.

Perhaps that Least Water-lily I picked lies too lightly on my conscience, almost as lightly as it floated on the loch, but it is not so with a more precious plant. I was staying at the railway station of Blair Atholl, not in the station itself, but in a railwayman's house that overlooked the line. The shining rails, that carried trains to the north, carried my thoughts to the Sow of Atholl, the one mountain home of Scottish Menziesia. My friend, Fordie Forrester, was staying at Blair Atholl too, and I reflected that he had a car that could take me to the mountain-foot. That he had no special interest in plants made me regard his mind as a kind of virgin soil in which to sow a passion to see Scottish Menziesia. Sure enough, I had not talked to him for ten minutes when I felt I had never known a keener botanist. 'But we shall not find it', I said; 'I am not sure it is not extinct'. 'Don't you worry; we shall find it', he replied; 'we start at ten o'clock tomorrow morning'. So next day we set off with our wives, Fordie wearing his kilt for the occasion. Everything looked bright and cheerful in the rain, Devil's-bit and Marsh Lousewort, the coloured patches of Sphagnum-moss, even the Cotton-grass in the black bogs, tossing their wool-tegged distaffs. A low mist halved the mountains; of the Sow itself only the lower parts were seen, its head wrapt in heavenly contemplation. 'Stop here', I cried at last, and getting out of the car we stood and viewed what was left of the mountain by the mist, which appeared to be always rising but never rose. 'We must spread out', I explained; so we began to climb in parallel lines that were soon lost in the grey darkness. For a time we wandered about, suddenly appearing to one another and disappearing, like the four lovers in *The Midsummer Night's Dream*. That I was the one to find the plant was not surprising, for I had received directions which I had not thought necessary to impart to the others. But having found it, how was I to find them? I kept calling till I began to think I had lost them for ever, and even Scottish Menziesia seemed scarcely to compensate for the loss. But one by one they loomed in sight. As we stood round the plant I remembered the old man who had thanked me sarcastically for the common flowers, but had missed the Snow Gentian. 'I will show him', I said, and I stooped and picked a flower. That night it lay in a glass of water by my bed-side, taking its last drink on earth. In the morning, resolved that this plant should not go astray, I wrote a letter to tell him what I was sending. Then, taking it from the glass, I looked at it; but as I looked, a sense of shame that I had picked it swept over me. I felt I had to get rid of it at once, and, stepping quickly to the open window, I flung it out. I watched it fall into an empty truck standing in the station. Wherever that truck went, it did not take with it the Scottish Menziesia, at least not altogether; it lies too heavy on my conscience.

> ANDREW YOUNG, *A Prospect of Flowers*, 1945. A Minister of the Free Church of Scotland and distinguished poet, Andrew Young was careful to draw a distinction between botanists and what he called 'botanophils'

One morning, halfway through that first summer term at St. Ethelbert's, a small parcel arrived for me. I undid it, and with difficulty choked down my tears. It contained orchids found by my old Nurse near the cottage, whither the family had already repaired for the summer. Not for years—not till I had left school—should I ever be able to find these orchids myself again: the Lady, the Green Man, the Early Spider—none of them grew near St. Ethelbert's. For the months of May, June and July I was condemned, for what seemed all eternity (and at that age, there is little difference between five years and eternity) to an unhappy exile, if not in a flowerless, at least in an orchidless world. I realized it for the first time, that morning; and the yellowish spikes of the Man, the purple-spotted pagodas of the Lady, awoke in me a nostalgia which was no ordinary homesickness, but a sense of greater loss. I realized, at last, that my childhood was nearly over.

Hurrying to be in time for Prayers (with the dread word 'Honour' echoing in my ears) I stuffed the orchids into my tooth-glass in the dormitory—as usual, with a sense of shame and embarrassment, rather as a priest in Russia might prepare to celebrate Mass before an assembly of keen party-members . . . Yet, like the priest, I was secretly assured of the validity of the rite; privately I despised my tittering companions, recognizing the shallowness of their school-bred, conventional enthusiasms, knowing them incapable, in nearly every case, of even a moment's genuine emotion.

I hurried down to Prayers; but not before I had noticed, among the other orchids, an unfamiliar one: pale pink, almost white, with a crimson lip narrowly divided into four tendril-like lobes . . . Later in the day, I unpacked Edward Step from my play-box, and once again turned up the passage about *Orchis militaris*. Had I found it at last—the Military Orchid? Again the annihilating, impotent nostalgia swept over me: the orchid had been found quite near our cottage, in the park in which I, myself, had found the Green Man, the Bee, the Pyramidal, for many a summer . . . And now, in the very first year of my absence, this distinguished stranger had elected to turn up there. Was it the Military? Once again, as when Mr. Bundock had brought me the Lady, the old conflict was revived: I wanted my plant to be the Military, and knew that, by accepting Colonel Mackenzie as my authority, I was justified in so calling it. But alas! there was that qualifying clause in Edward Step about 'a sub-species known as the Monkey Orchid . . . with narrower divisions of the crimson lip . . . occurring in the same counties as the type, with the addition of Kent.' I looked at the plant again: the divisions of the lip could not well have been narrower; moreover, they were indubitably crimson. And the plant had been found in Kent . . . Reluctantly I decided that it was, after all, the 'sub-species known as the Monkey Orchid'. (Perhaps St. Ethelbert's had already purged me of that mental dishonesty which had enabled me, years before, to label Mr. Bundock's orchid the Military.)

So the Military was, after all, still unfound. It was nice to have the Monkey; but there seemed to me something slightly inferior about a 'sub-species'. The very phrases of Edward Step sounded faintly derogatory—'a sub-species *known* as the Monkey Orchid' . . . it suggested the subtly insulting phraseology of the police-court: 'a woman *described* as an actress'. I could not know that my old Nurse's 'find' was to prove one of the more important plant-records of the century.

A year or two later, the season was early, and the summer term must have started late; walking across the Park, on the last day of the Easter holidays, I found a single plant of the Monkey Orchid. By that time, I had a better idea of its importance: I had discovered that it was promoted, nowadays, to specific status, and could be ranked with the Military on equal terms. Later, I heard that one or two other people had found it in the same place at about this period: a lady staying in our village had drawn it, and the drawing was hung—and still hangs—in the Canterbury Museum. But after 1923—when I found it myself—it seems to have disappeared entirely from the district. It was not till many years later that I realized how portentous its appearance there had been.

Colonel Godfrey, I found, in his immense and erudite *Monograph* on the British Orchids, could quote only one record for the Monkey Orchid in Kent—it had, apparently, been found, in the early nineteenth-century, by the Rev. S. L. Jacobs, near Dover, and never re-discovered since. Colonel Godfrey, indeed was sceptical even of this single appearance, attributing it to an error of identification, or to a windblown seed.

I wrote to him, enclosing a floret of a pressed specimen, which had fortunately survived the years. He agreed that it was undoubtedly the Monkey Orchid. And so, in the warm, rainy days of Munich I wrote the story of the Kentish Monkey for the *Journal of Botany*: remembering that morning at St. Ethelbert's—the tittering boys, the sneers of Mr. Wilcox, the hustle to be ready in time for Prayers; and the delicate, aristocratic flower, one of the rarest and most beautiful in the British Flora, stuffed hurriedly and ignominiously into a tooth-glass.

The other day, I made the pilgrimage to Oxfordshire to visit the Monkey in its sole remaining locality: a chalky hillside overlooking the river, within too-easy reach of a popular boating-resort. There it was: half-hidden among the rough grasses, smaller than its Kentish fellows, but the same charming and exquisitely-formed flower that I had stuffed into my toothglass at St. Ethelbert's some twenty-five years before. The botanist who accompanied me pointed out that I was possibly the only living person who had seen the Monkey Orchid growing in two separate British localities. If so, it is perhaps a small claim to fame: but one of which I am extremely proud.

JOCELYN BROOKE, *The Military Orchid*, 1948

And here, suddenly, a hundred miles and twenty years away from it, there is a change even while one is looking back. It is almost May, and suddenly on a late April morning the wind changes, north to west and west to south, and the sky is softened into a sea of white islands, the sun breaks hot, and the wood, as though reflecting that warmth and that whiteness, breaks suddenly into columns of shining smoke. The wild cherry is in bloom.

For about four months, from March until the end of June, the wild flowering trees of this country are at their best. Unlike the trees of gardens, they seem to have no years of shyness, the uncertainties of cold and rain and sunshine never seem to affect them,

and with one or two distinguished exceptions they flower only at
that time, between first spring and midsummer. Yet while they
flower they are immeasurably glorious. The best of them are the
trees of poets; they are to the world of trees what the lily and
the daffodil are to the world of flowers. The humblest of them
are the treasure grounds of bees, the ivy in late summer as rich
with bloom and honey-scent as the sloe is thick with scentless
stars of snow in March. And all are common trees, hedgeside
and wayside trees for the most part, with nothing exclusive or
niggardly or exotic about them. They are the friendliest and love-
liest of trees. The blackthorn opens the season and the honeysuckle, I suppose, ends it. And they stand distant not only in
time but in almost all other respects, in scent and shape and
colour and effect, the blackthorn so very cold and snowy, the little star-shaped blossoms so pure and icy, the real pristine emblems
of the breaking spring, and then the honeysuckle rich and sun-
coloured, the flower-head a lovely and fantastic clustering of many
flowers in one, a cornucopia of softest amber and cream and
ruby, with the scent of heaven. And if the blackthorn is one of
the shortest and perhaps even the very shortest of all in its season of flowering, then the honeysuckle is certainly the longest. It
begins in midsummer and goes on through haytime and harvest,
renewing itself in warm autumns until that richness of wine and
amber is lost among the colours of dying leaves about the empty
corn-fields. The Irish, I think it is, have a legend that the honeysuckle is the strongest of trees. They might with equal truth have
had a legend that it was the tree that never rested. For the flowers
have scarcely been replaced by the shining cherry-coloured seeds
before the vine is breaking into new leaf again, so that often in
midwinter the honeysuckle is the true evergreen of the woods, in
brilliant and almost full leaf long before the black branches of
the sloe have been threaded by the flower-buds of cream that are
its first signs of life in March.

The blackthorn blooms on into April, reaching its glory as the
cherry begins. By the end of the month the cherry out-flowers
it. There is something earthy about the blackthorn; it is a dwarfish
tree, almost stunted, always near the earth. But the wild cherry
flowers against sky, in white grace and magnificence, with true
ethereal loveliness, visible from afar off. In orchards the cherry
will grow to great extent, but not height. But in woods the wild
cherry, hemmed in by oaks and chestnuts and trees of equal

growth, rarely grows to great extent, but very often to immense height. A wild cherry will grow to seventy feet, flowerless until the extreme tip lifts itself above the crowd of neighbouring trees, the thick white clustering of blossom floating above the wood like a cloud on the mountain of colouring branches, never still in an April wind. After the catkins, it is the first glory of the woods. It is equalled only by the hawthorn, the may, the first glory of the hedges. The may is erratic; of all the wild flowering trees it fluctuates most with the season. The cherry blooms infallibly in April, but the coming of the may is never certain. Often on May Day it would be hard to find the traditional branch of it, though I have seen it in bloom in other and colder springs in the first weeks of April. But when it finally blossoms there is no uncertainty at all about it; its flowers are the risen cream of all the milkiness of May-time. Its scent has the exotic heaviness of summer in it, very like the pungent vanilla half-sweetness of meadow-sweet. It is so like the blackthorn and yet so unlike it; the blackthorn, with its black naked twigs that have no suppleness or tenderness, bears flowers of frost, but the may branches are never cold or stiff or naked. The leaves of emerald are full blown and the flowers with their pin-hearts of claret spill and foam and cascade down the hedgesides with a summery richness that no other English tree, not even the elder, can equal, splashing the grass and the earth underneath them with cream that turns to pink as time goes on and the sun increases.

The crab comes with the may, and the elder after them. The crab stands apart. It is the sweetest of all trees, the pure cups of pink and white truly sweet, without the vanilla drowsiness of the rest, the upturned blossoms smooth and light and shining, like spring silk. And after it the elder, bringing back the odours of may and meadow-sweet again, only half-sweet, falls again like the may in great cascades of even richer cream.

How is it that this current of cream and white and pink goes on and on through the wild trees of England almost without break or variation? The chestnut and the crab and the wild rose and even the blackberry are white and pink. The dogwood and the elder and the lime are cream. The rest are white. And all are scented, either with that summery faintness of the may or with the absolutely pure sweetness of the crab and the chestnut and the rose. We have no wild exotic blossoming trees of scarlet or blue or purple. There is a sort of northern delicacy, almost fragility, about them all. The flames of the chestnut candelabra burn sweetly and quietly, many little flames of softest pink in the white cups of wax above the drooping clusters of seven leaves. The rose has no passion, only that immeasurable and matchless sweetness that fills the hot days of June and July as the heavenliness of the lime drenches the summer nights.

Gorse and broom alone break the sequence of pink and cream and white. And curiously they are the smallest and most brilliant of all. They are trees of flame, the broom flaming up in May with little passionate tongues of yellow, the gorse burning throughout the year from one end to another, flickering or flaming up with solitary or countless flames of blossom according to the season, never resting or going out, a tree of perpetual flowering fire and darkness.

H. E. Bates, Through the Woods, 1936

Our habit in the spring—or rather in the unvegetative beginning of the year when we suspect that spring might, with luck, begin to happen in a few weeks' time—is to pick branches and put them in water, and wait for their leaves or flowers to open.

In this how conservative we are. We stick to a very few kinds; which are not all of them the ones that look best in a room.

I am all for respecting the tradition of plants, as well as for knowing about them historically, and scientifically—up to a point beyond which the amateur does not need to go; but sometimes we should trust more to our own senses of shape, design, and colour and not stuff vases only with hazel, sallow, and rumbustious girders of horse chestnut.

Elm: Nothing in a room makes more of a remarkable net of twinkling shapes than elm twigs, broken off when they are in bud. No branches fit themselves more easily, and *firmly*, into a jug.

Ash: In a lumpier way, I am also for arrangements of black ash buds on gnarled and twisting grey twigs. They open from black to pink, or purple-pink; which gives way to a peculiar green, till at the end of your twigs you have little crisp bushes like the carrageen seaweed in miniature; or like hydroids from a sea pool, enlarged.

*

Going abroad does good—or harm—to our conceit about 'English' plants. Suddenly the rarities we are so proud to know, so glad to find at home—are not rare at all. Boys by the road hold out bunches of Lily of the Valley picked from woods where they are as thick as wood anemones; fields are blue with Cornflower (proper Cornflower, not Scabious), banks are particoloured or multicoloured with the plant which the French carelessly call *Queue de Renard*, Fox Tail, and which at home—the Purple Cow-wheat or Field Cow-wheat, or *Melampyrum arvense*—is one of the scarcest, most scattered prizes of secret knowledge and secret contemplation. Have a picnic in northern Spain, along the Cantabrian Sea and under the Picos de Europa, in a grove of old chestnut trees, and you tread, you light the stove, you kneel, on the finest plants of Acrid Lobelia—what a name!—the *Lobelia urens* which just manages to struggle into our western counties. Where it was 'first found', as we say, on a great Devonshire hillside, and where it still grows, in this chief of its English stations—in a hen run—there stands a Lobelia Cottage. Plant devotees arrive as though on pilgrimage, and contemplate, and do not dare to pick.

Going abroad can jolt our ideas of the proper keepings or whereabouts of some kinds of plant. Once I stopped to make coffee in mid morning in a valley of the Alps below the slightly horrible or fearful pass of Mt Cenis. It was autumn, and among the shingle and boulders brought down and spread wide by a stream there glittered millions of orange berries, through grey leaves, on shrubs which my fingers discovered to be prickly to a savage extreme.

How surprised I was when the shrub—after some rapid recourse to a French flora—turned out to be that scarce plant of the fixed sand-dunes of the North Sea which in our insular way we call 'Sea Buckthorn', though in fact it is more aberrant, less in place, along the shore than along streams in the Alps.

'In stony or sandy places, especially in beds of rivers and torrents, in central and eastern Europe and central and Russian Asia,' says a flora. '*Also occasionally near the seacoasts . . .*'

And when we come home, after such corrective, and so to say, deinsulating discoveries?

Oddly the old habit of mind reasserts itself; slowly the old aura of scarcity, rarity, botanical self-satisfaction, does return. I still make an annual pilgrimage in February to a valley where about fifty plants of Stinking Hellebore show their mineral-green flowers above the February snow, though I have seen plants of it by the

million on southern limestones. I still go in May to a scraggy
hillside off the Cotswolds which has two or three acres of wild,
rare, Lilies of the Valley.

GEOFFREY GRIGSON, 'Against Catkins' and 'Going Abroad',
A Herbal of All Sorts, 1959

There are three well-known myths about hedges. People suppose
that they are specially English, or British; that they are very
artificial, the mere result of somebody planting a hedge, and do
not have a life of their own like woods or heaths; and that they
are all, or nearly all, a mere 200 years or so old.

The first myth is soon disposed of. Western Normandy is *bocage*
country, the tangle of intricately and massively hedged fields
that complicated the fighting of World War II. In sudden contrast are the wide-open hedgeless plains of east Normandy. Most
of France is a patchwork of hedged *bocage* areas and hedgeless
champagne areas. The distinction continues far into Europe. There
are hedged areas in northern Italy, the Austrian Alps, and around
Mount Helicon in Greece; even Crete has one or two hedged
parishes. In America I have seen thousands of miles of hedge in
a dozen States from Vermont to Texas; there are hedged fields
in the Peruvian Andes. Regions with and without hedges are to
be found over much of the world. In general, traditions of hedgeless open-field or prairie-farming belong in great plains or broad
valleys. Where the whole of a region is not hedged, hedges
tend to go either with hilly terrain or with the neighbourhood of
woods.

Hedges are of many kinds. In Cornwall a hedge is an earth or
stone bank, which may be—but often is not—topped with bushes
or trees. In Michigan a 'fence-row' consists of biggish trees closely set on a low bank of loose stones. Neither of these needs to
be kept in being by regular cutting or plashing as does the classic
English Midland hedge. Historical documents are sometimes vague
about the distinction between hedges and fences: 'fence' in Enclosure Acts can mean any field boundary. Hedgerows are occasionally confused with narrow woods: a map and survey of
Plumberow in Hockley (south-east Essex) in 1579 shows a curious network of fields separated by 'hedgerowes' some of which
were 100 yards thick.

WHERE HEDGES COME FROM

In Britain, hedge-planting is familiar and well documented; nearly all more recent hedges have certainly been planted. Let us not slip into the generalization that all hedges have been planted: there are two other ways to get a hedge.

North America lacks this hedge-planting tradition: settlers fenced their fields with wood or wire. Yet the United States now has more miles of hedge than Great Britain. Americans believe that nearly all their hedges arose by default. Tree saplings sprang up alongside the fences and eventually replaced them. I have studied stages of the process. The prairies of middle Texas, near Waco, originally maintained by wild animals, were parcelled out into farms and fields by barbed-wire fences in the 1880s. Seedlings of Texas elm, black oak, Texas ash, prairie sumach, poison-ivy, and many other trees and shrubs have sprung up at the bases of the fences, which have sheltered them from browsing and cultivation. The hedges have advanced gradually: aerial photographs prove that many of them were discontinuous, or not there at all, twenty years ago. There can be no question of any planting—this has been a time of declining prosperity. Tree seeds have arrived naturally from the wooded canyons nearby. People have failed to prevent the trees from growing, and doubtless have found them a useful relief from replacing rotten fence-posts.

Later stages can be seen in other States. Michigan is parcelled out by nineteenth-century fence-rows. In Massachusetts the seventeenth-century fences were replaced by eighteenth-century hedges, which now take the form of rows of old trees through the nineteenth-century secondary woods which have engulfed the fields. American hedges are usually within three miles of woodland. They are seldom managed except by casual woodcutting. They are nearly always mixed, of at least five species, whatever their age.

Has this happened in Britain? Our tree species have much the same colonizing powers as their American sisters. Close to Hayley Wood (Cambridgeshire) there was a railway from 1862 to 1969, separated from the adjoining field by the usual wire fence and shallow ditch. Since the railway has been disused, trees and shrubs rooted at the base of the fence have grown into an almost continuous row which it will be possible to maintain as a hedge. It is a mixed hedge of hazel, hawthorn, ash, briar, blackthorn, maple, etc., and differs from the oak, birch, and sallow which have sprung up on the disused railway itself. By the usual criterion of hedge-dating it ought to be 700–800 years old. It cannot be older than 1862; the abundant hazel suggests that it began to develop before 1930, when grey squirrels arrived in the area. Probably the railwaymen who mowed the grass verges were unable to reach a narrow strip at the base of the fence.

That this has not happened more often in the last 200 years is because farmers and labourers have had time on their hands in slack periods and have chosen to spend it in tidying, 'brushing', and suppressing young trees. This may not always have been so. The American parallel suggests that hedges arise wherever men are few and acres many, and especially in times of moderate agricultural depression. Whenever a ditch, bank, lynchet, or earthen wall is neglected for a few years not too far from a source of tree seed a hedge will result. Fences turn into hedges by birds sitting and dropping seeds; the fence protects the incipient hedge. When prosperity returns the young trees will be managed as a hedge and their origin will be forgotten.

Hedges arise in a third way as the *ghosts* of woods that have been grubbed out leaving their edges as field boundaries. The marginal trees, often already forming a hedge to protect the wood's interior, may be left as a hedge having woodland, rather than hedgerow, characteristics. At Shelley (Suffolk), I have been shown a remarkable roadside hedge 600 yards long composed almost entirely of the pry tree (small-leaved lime) with occasional service. Pry and service are woodland, not hedgerow trees; pry is the commonest tree of ancient woods in the area but is unheard-of growing by hundreds in a hedge. The mystery is resolved by an eighteenth-century map which shows 'Withers Wood' adjoining just that length of road where the pry hedge now stands. Judith's Hedge, said to have been made round Monks' Wood (Huntingdonshire) in *c.*1080 by one Countess Judith is now in part detached from the wood owing to grubbing in the seventeenth century.

Oliver Rackham, *The History of the Countryside*, 1980

After a simple lunch we set out in Gilberto's boat crossing a small *paraná* to the *igapó* where we found the buds. On the way we stopped only once, when we sighted more reddish cactus leaves, but they were of a phyllocactus, so we continued. We arrived at the *igapó* at four in the afternoon. The two buds which had been tightly folded the previous day had burst, their twisted petals now loosely closing the flowers, just as I had seen with Sue on my journey six years before. Another flower was drooping and two more were ready to open, surely it had to be that night. We decided to stay, and while Gilberto helped me to the roof Sue and Sally took a dip in the dark water.

The plant was obviously healthy, its leaves flattened to the trunk were green turning to crimson and some were completely changed. From the vantage point I observed that there were many epiphytes on the tree including a gesneriad which partly masked the cactus. From the midst of the foliage a tiny lizard, green with crimson markings was watching me. He was beautifully camouflaged and blended perfectly with the cactus leaves. I kept still, the warmth of the low afternoon sun on my back until, his suspicion satisfied, the sleek creature scuttled along one leaf and cleverly dropped to another. The trunk was a wonderful home with all the roots of the epiphytes packed with tiny plant remains forming a humus for a complete garden of species. Ants busily foraged and a host of other insects, beetles among them, scurried about. The lizard no longer bothered by my presence simply darted from one good feast to another.

Sue and Sally scrambled aboard just as the sun was setting. It sent a golden shimmer across the water, lighting the forest a pale yellow green. The birds were quieter and slowly, one by one, frogs joined in a resounding chorus. We prepared for our vigil. Gilberto produced some torches and we ate the last of our dry biscuits from Manaus with coffee brewed by Mariá. Within the hour the sky was black and only the brightest stars showed through a layer of thin cloud covering the silent forest. We took turns to keep watch on the buds using a torch beam briefly, so we would not disturb the opening. Once in the confusion of the dark a torch slipped over the side of the boat and we watched its powerful beam slowly dim as it fell through the deep tea-coloured water.

THE FLOWERING

We had been waiting almost two hours and the buds had not changed. The flooded forest was quiet apart from the frogs and an occasional grunting or whistling call from a night prowling animal—all the inhabitants can swim or climb.

As I stood there with the dim outline of the forest all around I was spellbound. Then the first petal began to move and then another as the flower burst into life. It was opening so quickly. A chair was passed to the roof, and Sally found the rest of my sketching materials whilst Sue stood beside me with a small portable battery light. Brian and Tony prepared to record the event. The flower was nearly open but the strong photographic lights seemed to slow down the opening, so I begged that the light be dimmed, and we continued with only a faint illumination and the light of the full moon rising over the darkened rim of forest.

In the early stages an extraordinary sweet perfume wafted from the flower, and we were all transfixed by the beauty of the delicate and unexpectedly large bloom, fully open in an hour.

The petals or perianth leaves were long and slightly wider than the outer ones we had seen on the closed flowers. Open at last it was a curious flower and totally different to any I had painted before. Also it was a challenge as I was balanced precariously on the six foot wide top of the boat and anxious that the slightest movement would cause us to tip. Luckily the night was still and without a breath of wind, so I could work easily. As I sketched I hoped for the pollinator to arrive, which specialists deduce from the structure of the flower should be either a large moth or a bat, and possibly quite specific to this species. Very few accounts have been made about pollination in the wild for night-flowering cacti, so we were hoping for a discovery. But no large insect came by and the only nocturnal visitors we saw were dozens of small flies which gathered on the stamens. The strong perfume faded after two hours though it was still possible to detect it by bending close to the flower.

The moon climbed higher disappearing occasionally behind cloud though for most of the night giving a bright clear view of the *igapó* dotted with the pale grey trunks of the weather sculpted trees. Our vigil was long and I conclude that our intrusion had deterred the pollinator, upsetting the delicate balance between the plant and the animal which has taken tens of millions of years to evolve. As the Moonflower began to change again, the effect of intrusion multiplied across all the species in the length and breadth of Amazonia was too much to contemplate. With the

dawn the flower closed and we watched fascinated and humbled
by the experience.

> MARGARET MEE, *Flowers of the Amazon Forests*, 1988. Margaret
> Mee, like many adventurous botanical artists before her, had
> a vivid and attentive writing style, and a reverence for the
> landscapes and species she painted

In March, renewal of the southeast monsoon brings the long
rains. Rains vary from region to region, according to the winds,
and since the winds are not dependable, seasons in East Africa
have a general pattern but can seldom be closely predicted. The
cyclical wet years and years of drought are a faint echo of the
pluvials and interpluvials of the Pleistocene. In 1961 drought had
destroyed thousands of animals; the next year floods killed thousands more. In the winter of 1969 rain fell in the Serengeti almost
daily, and no one knew whether the short rains of late autumn
had failed to end or the long rains of spring had begun too early.
A somber light refracted from the water gleamed in the depressions, and the treeless distances with their animal silhouettes, the
glow of bright flowers underfoot, recalled the tundras of the north
to which the migrant plovers on the plain would soon return.

The animals had slowed, and some stood still. In this light
those without movement looked enormous, the archetypal animal cast in stone. The ostrich, too, is huge on the horizon, and
the kori bustard is the heaviest of all flying birds on earth.
Everywhere the clouds were crossed by giant birds in their slow
circles, like winged reptiles on an antediluvian sky.

One morning the dog pile broke apart before daylight and
headed off toward the herds under Naabi Hill. Unlike lions, which
often go hungry, the wild dogs rarely fail to make a kill, and this
time they were followed from the start by three hyenas that had
waited near the den. The three humped along behind the pack,
and one of the dogs paused to sniff noses with a hyena by way
of greeting. In the distance, zebras yelped like dogs, and the dogs
chittered quietly like birds as they loped along. As the sun rose
out of the Gol Mountains, they faked an attack on a string of
wildebeest and moved on.

A mile and a half east of the den, the pack cut off a herd of
zebra and ran it in tight circles. There were foals in this herd,
but the dogs had singled out a pregnant mare. When the herd
scattered, they closed in, streaming along in the early light, and

almost immediately she fell behind and then gave up, standing motionless as one dog seized her nose and others ripped at her pregnant belly and others piled up under her tail to get at her entrails at the anus, surging at her with such force that the flesh of her uplifted quarters quaked in the striped skin. Perhaps in shock, their quarry shares the detachment of the dogs, which attack it peaceably, ears forward, with no slightest sign of snapping or snarling. The mare seemed entirely docile, unafraid, as if she had run as she had been hunted, out of instinct, and without emotion: only rarely will a herd animal attempt to defend itself with the hooves and teeth used so effectively in battles with its own kind, though such resistance might well spare its life. The zebra still stood a full half-minute after her guts had been snatched out, then sagged down dead. Her unborn colt was dragged into the clear and snapped apart off to one side.

The morning was silent but for the wet sound of eating; a Caspian plover and a band of sand grouse picked at the mute prairie. The three hyenas stood in wait, and two others appeared after the kill. One snatched a scrap and ran with it; the meat, black with blood and mud, dragged on the ground. Chased by the rest, the hyena made a shrill sound like a pig squeal. When their spirit is up, hyenas will take on a lion, and if they chose, could bite a wild dog in half, but in daylight, they seem ill at ease; they were scattered by one tawny eagle, which took over the first piece of meat abandoned by the dogs. The last dog to leave, having finished with the fetus, drove the hyenas off the carcass of the mare on its way past, then frisked on home. In a day and a night, when lions and hyenas, vultures and marabous, jackals, eagles, ants, and beetles have all finished, there will be no sign but the stained pressed grass that a death ever took place . . .

Another day, by a korongo, I helped Schaller collect a dying lioness. She had emaciated hindquarters and the staggers, and at our approach, she reeled to her feet, then fell. In the interests of science as well as mercy, for he wished an autopsy, George shot her with an overdose of tranquilizer. Although she twitched when the needle struck, and did not rise, she got up after a few minutes and weaved a few feet more and fell again as if defeated by the obstacle of the korongo, where frogs trilled in oblivion of unfrogly things. I had the strong feeling that the lioness, sensing death, had risen to escape it, like the vultures I had heard of somewhere that flew up from the poisoned meat set out for lions, circling higher and higher into the sky, only to fall like stones as life foresook them. A moment later, her head rose up, then flopped for the last time, but she would not die. Sprinkled with hopping lion flies and the fat ticks that in lions are a sign of poor condition, she lay there in a light rain, her gaunt flanks twitching.

The episode taught me something about George Schaller, who is single-minded, not easy to know. George is a stern pragmatist, unable to muster up much grace in the face of unscientific attitudes; he takes a hard-eyed look at almost everything. Yet at this moment his boyish face was openly upset, more upset than I had ever thought to see him. The death of the lioness was painless, far better than being found by the hyenas, but it was going on too long; twice he returned to the Land Rover for additional dosage. We stood there in a kind of vigil, feeling more and more depressed, and the end, when it came at last, was shocking. The poor beast, her life going, began to twitch and tremble. With a little grunt, she turned onto her back and lifted her hind legs into the air. Still grunting, she licked passionately at the grass, and her haunches shuddered in long spasms, and this last abandon shattered the detachment I had felt until that moment. I was swept by a wave of feeling, then a pang so sharp that, for a moment, I felt sick, as if all the waste and loss in life, the harm one brings to oneself and others, had been drawn to a point in this lonely passage between light and darkness.

PETER MATTHIESSEN, *The Tree where Man Was Born*, 1972

I saw my first peregrine on a December day at the estuary ten years ago. The sun reddened out of the white river mist, fields glittered with rime, boats were encrusted with it; only the gently lapping water moved freely and shone. I went along the high river-wall towards the sea. The stiff crackling white grass became limp and wet as the sun rose through a clear sky into dazzling mist. Frost stayed all day in shaded places, the sun was warm, there was no wind.

I rested at the foot of the wall and watched dunlin feeding at the tide-line. Suddenly they flew upstream, and hundreds of finches fluttered overhead, whirling away with a 'hurr' of desperate wings. Too slowly it came to me that something was happening which I ought not to miss. I scrambled up, and saw that the stunted hawthorns on the inland slope of the wall were full of fieldfares. Their sharp bills pointed to the north-east, and they clacked and spluttered in alarm. I followed their point, and saw a falcon flying towards me. It veered to the right, and passed inland. It was like a kestrel, but bigger and yellower, with a more bullet-shaped head, longer wings, and greater zest and buoyancy of flight. It did not glide till it saw starlings feeding in stubble, then it swept down and was hidden among them as they rose. A minute later it rushed overhead and was gone in a breath into the sunlit mist. It was flying much higher than before, flinging and darting forwards, with its sharp wings angled back and flicking like a snipe's.

This was my first peregrine. I have seen many since then, but none has excelled it for speed and fire of spirit. For ten years I spent all my winters searching for that restless brilliance, for the sudden passion and violence that peregrines flush from the sky. For ten years I have been looking upward for that cloud-biting anchor shape, that crossbow flinging through the air. The eye becomes insatiable for hawks. It clicks towards them with ecstatic fury, just as the hawk's eye swings and dilates to the luring food-shapes of gulls and pigeons.

To be recognised and accepted by a peregrine you must wear the same clothes, travel by the same way, perform actions in the same order. Like all birds, it fears the unpredictable. Enter and leave the same fields at the same time each day, soothe the hawk from its wildness by a ritual of behaviour as invariable as its own. Hood the glare of the eyes, hide the white tremor of the hands, shade the stark reflecting face, assume the stillness of a tree. A peregrine fears nothing he can see clearly and far off. Approach him across open ground with a steady unfaltering movement. Let your shape grow in size but do not alter its outline. Never hide yourself unless concealment is complete. Be alone. Shun the furtive oddity of man, cringe from the hostile eyes of farms. Learn to fear. To share fear is the greatest bond of all. The hunter must become the thing he hunts. What is, is now, must have the quivering intensity of an arrow thudding into a tree. Yesterday is dim and monochrome. A week ago you were not born. Persist, endure, follow, watch . . .

(*December 15th.*) The warm west gale heaved and thundered across the flat river plain, crashed and threshed high its crests of airy spray against the black breakwater of the wooded ridge. The stark horizon, fringing the far edges of the wind, was still and silent. Its clear serenity moved back before me; a mirage of elms and oaks and cedars, farms and houses, churches, and pylons silver-webbed like swords.

At eleven o'clock the tiercel peregrine flew steeply up above the river, arching and shrugging his wings into the gale, dark on the grey clouds racing over. Wild peregrines love the wind, as otters love water. It is their element. Only within it do they truly live. All wild peregrines I have seen have flown longer and higher and further in a gale than at any other time. They avoid it only when bathing or sleeping. The tiercel glided at two hundred feet, spread his wings and tail upon the billowing air, and turned down wind in a long and sweeping curve. Quickly his circles stretched away to the east, blown out elliptically by the force of the wind. Hundreds of birds rose beneath him. The most exciting thing about a hawk is the way in which it can create life from the still earth by conjuring flocks of birds into the air. All the feeding gulls and lapwings and woodpigeons went up from the big field between the road and the brook as the hawk circled above them. The farm seemed to be hidden by a sheet of white water, so close together were the rising gulls. Dark through the white gulls the sharp hawk dropped, shattering them apart like flinging white foam. When I lowered my binoculars, I saw that the birds around me had also been watching the hawk. In bushes and trees there were many sparrows, starlings, blackbirds and thrushes, looking east and steadily chattering and scolding. And all the way along the lanes, as I hurried east, there were huddles of small birds lining the hedges, shrilling their warning to the empty sky.

As I passed the farm, a flock of golden plover went up like a puff of gunsmoke. The whole flock streamed low, then slowly rose, like a single golden wing. When I reached ford lane, the trees by the pond were full of woodpigeons. None moved when I walked past them, but from the last tree of the line the peregrine flew up into the wind and circled east. The pigeons immediately left the trees, where they had been comparatively safe, and flew towards North Wood. They passed below the peregrine. He could have stooped at them if he had wished to do so. Woodpigeons are very fast and wary, but like teal they sometimes have a fatal weakness for flying towards danger instead of away from it.

In long arcs and tangents the hawk drifted slowly higher. From five hundred feet above the brook, without warning, he suddenly fell. He simply stopped, flung his wings up, dived vertically down. He seemed to split in two, his body shooting off like an arrow from the tight-strung bow of his wings. There was an unholy impetus in his falling, as though he had been hurled from the sky. It was hard to believe, afterwards, that it had happened at all. The best stoops are always like that, and they often miss. A few seconds later the hawk flew up from the brook and resumed his eastward circling, moving higher over the dark woods and orchards till he was lost to sight. I searched the fields, but found no kill. Woodpigeons in hawthorns, and snipe in the marshy ground, were tamed by their greater fear of the hawk. They did not fly when I went near them. Partridges crouched together in the longest grass.

Rain began, and the peregrine returned to the brook. He flew from an elm near the bridge, and I lost him at once in the hiss and shine of rain and the wet shuddering of the wind. He looked thin and keen, and very wild. When the rain stopped, the wind roared into frenzy. It was hard to stand still in the open, and I kept to the lee of the trees. At half past two the peregrine swung up into the eastern sky. He climbed vertically upward, like a salmon leaping in the great waves of air that broke against the cliff of South Wood. He dived to the trough of a wave, then rose steeply within it, flinging himself high in the air, on outstretched wings exultant. At five hundred feet he hung still, tail closed, wings curving far back with their tips almost touching the tip of his tail. He was stooping horizontally forward at the speed of the oncoming wind. He rocked and swayed and shuddered, close-hauled in a roaring sea of air, his furled wings whipping and plying like wet canvas. Suddenly he plunged to the north, curved over to the vertical stoop, flourished his wings high, shrank small, and fell.

He fell so fast, he fired so furiously from the sky to the dark wood below, that his black shape dimmed to grey air, hidden in a shining cloud of speed. He drew the sky about him as he fell. It was final. It was death. There was nothing more. There could be nothing more. Dusk came early. Through the almost dark, the fearful pigeons flew quietly down to roost above the feathered bloodstain in the woodland ride.

J. A. BAKER, The Peregrine, 1967

Thursday 22 December [1983]

Wild night with sleet showers, strong wind and thunder. Would the badgers be above ground (if they had escaped the diggers) in such weather? Sleet and thunder had stopped, however, by the time I reached the Ridge. Wind and wet on the overhead electricity cables was deafening. Water lying on the fields and running down the paths.

I stood on top of the Wheatfield Sett bank; visibility good yet no sign of my badgers. Called, but the wind tore the sound away. Walked along the top of the bank where they would most likely see movement—if still about. Stood above the beehives—wind, cold and dread gnawing at me. Heard a sharp bomp—another—and, glancing down, saw a hive at a curious angle, apparently on the move! Lesley at it again and this time helped by her mate. A pair of *very* muddy, breathless badgers having a marvellous dig-out. Sat amongst the wet nettles; dirty snouts exploring my face; earth-caked paws clambering over my lap; arms round two soaked little bodies—the relief indescribable!

Before going home, I tried to straighten the hive. Had no idea they were so heavy. Or that it had some occupants although, fortunately, very torpid! Eventually, got it level once more on its stone base, a mildly interested duo washing themselves near by . . .

Wednesday 28 December

Lovely night—mild, clear skies, crescent moon. I was musked by Lesley and her mate; the first time the young boar has done this. Plenty of food here for the badgers—everywhere very moist. Both *very* playful. Young Lesley appeared with her ball! Had forgotten all about it; what a bossy little character she is, but great fun.

No cars parked anywhere in or near these woods yesterday or this morning. I am still anxious about Monday's car and its occupants. Feel they were after the badgers, but have no proof of course. Will keep the car number [number entered in diary omitted here] and await developments.

Dawn is now about 7.15 a.m. (though later if it's cloudy). Standing on the Wildflower Path I watched the rising sun over Great Chantry Field gently transform the dead bracken next to me into a blaze of colour—literally setting it afire. The beauty of this place hurts . . .

Monday 2 January 1984

Badgers dug out an entrance and renewed bedding at the Old Cherry Sett, then went to earth there 5.55 a.m. Walked round the woods checking for cars and reached the top of Briarmead at first light, approx 7.20 a.m. It's been a mild, dry night with clear skies—tawnies very vocal an hour ago. Followed the wood edge to return once more to the Old Cherry Sett via the field above.

I was still some distance away when a man carrying a rifle appeared from the tree-cover at the path above the sett, and stood watching the lane. Thought at first that he was an early-morning bird-shooter—but if so, why the vigil? He turned and saw me walking towards him, looked back into the wood, then raised his rifle and fired at me. Stood watching him a moment, took a deep breath, and came on quickly just as he fired again. Felt sure he was both warning someone below him on the slope that I was around, and also trying to frighten me off—so I started to run in order to shorten the distance between us. He fired a third time, rather too near for comfort, then disappeared into the wood. Seemed an age reaching the path—the grass of the hayfield was long and wet. Sure I heard a dog bark once, then again, above the noise I was making through the bracken. Reached the Wildflower Path in time to see a car turn off the field on to Briarmead and disappear down the lane. Back as well as front seats occupied—three to four people. Much too far away to distinguish its number.

Returned to the sett to check the entrances. Dogs' prints in the fresh earth dug out by the badgers tonight, but otherwise undisturbed. Seems they got as far as sending the dogs down, then I came along. What can save the badgers when I return to work next Monday and can't be around so late? If only the Inspector would have his men patrol Briarmead two or three times a week at dawn as they did in April to August last year. There was no trouble here at all then. The sight of the police is a far better deterrent than I can ever be.

CHRIS FERRIS, *The Darkness Is Light Enough*, 1986

Gigi, unlike Melissa, will leave no descendants. Yet it would be difficult to overstate the extent to which this large sterile female has influenced the lives of the Kasakela chimpanzees, particularly the males. Since 1965 when she became sexually mature she has produced a new pink swelling, more or less regularly, every thirty days or so. Thus for more than twenty years she has been almost continually available to the Kasakela males for the gratification of their sexual desires. During that time her over-worked sex skin has swelled and shrivelled no less than two hundred and fifty times. By contrast, Fifi swelled only thirty times during the twenty-year period following *her* first pinkness. As a result of this repeated and unnatural stretching Gigi's swelling today is huge when compared with those of other Gombe females.

Right from the start Gigi radiated sex appeal. Time and again she has been the nucleus of large and excited sexual gatherings, surrounded by most of all of the males of her community. And once the adult males have gathered together, drawn by the magnetic presence of a sexually popular female, they are far more likely to move out to peripheral areas of their territory to patrol the boundaries. Thus Gigi's magnificent swelling has, time and again, served as a banner to rally the Kasakela males, encouraging them to perform valiant deeds in the protection and expansion of their territory.

In one respect Gigi's sexual popularity is hard to understand, since she often pulls away from her male partners before the completion of the sexual act. And she has been doing this for twenty-odd years. I assume that the males find such behaviour irritating, as well as frustrating, but it has never seemed to dampen their ardour. There are times, too, when Gigi is extremely reluctant to comply with the sexual demands of a male and on these occasions her suitors are often remarkably patient. I remember once when Figan was trying to mate with her. Gigi, who was reclining on the ground, her provocative swelling very much in evidence, totally ignored her suitor's vigorous shaking of branches. After a few moments Figan, his hair (among other things) fully erect, stood upright and swayed branches wildly above her recumbent form. Gigi, barely glancing at him, rolled over and lay on her back, staring at the canopy above. Nonplussed, Figan sat down for a moment, occasionally shaking a small branch in a jerky, irritated sort of way as, presumably, he wondered what to do next. Gradually his branching got more violent, his hair (if it were possible) bristled even more, and there was a wild gleam in his eye that, I thought, boded ill for Gigi if she continued to ignore him much longer. Apparently Gigi got the same message, for she suddenly rose, approached Figan and crouch-presented before him. But no sooner had he begun to copulate than she pulled away, screaming, and rushed off.

She then lay down again about ten yards from Figan, who stayed where she had left him. Presently he lay down too, and all was quiet for an hour. Then he approached Gigi again—and once more she utterly ignored his courtship. Not until he repeated his wild, branch-shaking swagger around her did she finally get up and crouch for him—but yet again she almost immediately pulled away and ran off. This time, tight-lipped and scowling, Figan followed and his courtship was a clear-cut threat. She responded quickly, but the outcome was the same. Except that Figan, thoroughly stimulated, finally completed the sexual act—into the air.

There can be no other Kasakela female who has been led off on so many consortships as Gigi. Time and again she has followed different males, usually reluctantly, to the various peripheral parts of the home range that they preferred. She has been, over the past twenty years, on forty-three such excursions that we know of: the figure is probably higher. In terms of evolutionary biology the males were 'wasting their time' since no measure of reproductive success for either partner could possibly result. However, the males did not know this, so they competed for her favours in all good faith. Moreover, there is little doubt in my mind that even if they *had* understood they would still have voted overwhelmingly in favour of the continued presence of Gigi in their midst.

In one other way Gigi has served the males of her community: she has helped the infants and juveniles to learn the ins and outs of the sexual act. Male chimpanzees are sexually very precocious. From the time they can totter, they show great interest in pink swellings, and they 'mate' pink females zealously throughout their childhood. Of course this is just practice—a male is unlikely to be able to father a child until he is between thirteen and fifteen years of age. But it sometimes seems that Gigi prefers the small sexual advances of infant and juvenile suitors to the more vigorous demands of the adult males. Often she crouches accommodatingly as soon as one of these youngsters starts to court her—approaching with his tiny erection and imperiously shaking a little twig. Indeed, she sometimes actively solicits the sexual attentions of youngsters. Once, for example, she suddenly went over to where Prof and Wilkie were playing a boisterous game, seized Prof by his elbow, pulled him from his playmate, and then, still maintaining her grip, crouched before him. Only when he complied with her wishes did she release him.

At other times she ignores these youngsters completely, however much they persist—and in such matters infants can be surprisingly single-minded for periods of half an hour or more. I remember one long journey when Gigi, fully swollen, was followed by three petulant juvenile suitors. Each of them was quietly whimpering to himself as he followed behind that tempting pink bottom. Each of them approached and shook branches every time she stopped. And each of them was utterly ignored by Gigi . . .

I think the males truly respect this tough and dauntless female, who has been such an integral member of their society for so long. And so, despite her idiosyncratic sexual behaviours, Gigi enjoys relaxed relationships with them and is a favourite grooming partner. Like the males she spends much time in noisy excited social gatherings, whereas most females, unless they are pink, prefer a more peaceful existence, choosing to spend days at a time with family members only and joining the larger groups merely for periodic spells of stimulation. Gigi, again like the males, spends a good deal of time in absolute solitude, whereas other females, after they have had their first baby (provided it lives) never experience real solitude again. For the rest of their lives they are always with one or more of their offspring. Having been a mother myself, I know full well that even a small baby can provide a sense of real companionship.

And so Gigi, in many ways, stands alone. For, despite her many male-like characteristics she is not a male: she has never been, and she never will be, fully integrated into the camaraderie of male society. Nor can she find companionship and comfort, as other females do, within a family. Of course she was part of a family once, but that was long ago. Even when I first knew her, when she was about eight years old, her only relative appeared to be the young male Willy Wally. And he moved away to the south with the Kahama males when the community divided.

With no infant of her own, no opportunity to create, for herself, that special group of close friends, a family unit, Gigi has instead cultivated a number of special relationships with a succession of infants. She became attracted to each of them when the infant was about one and a half years old—the age at which their mothers gave them relative freedom to interact with individuals outside the family circle. When she was with the family, and when the mother permitted it, Gigi would groom, play with or carry her current favourite. She helped to protect the infants too—she was particularly zealous in breaking up play sessions with older youngsters when they began to get too rough. In effect she assumed the role, for one infant after another, of the traditional maiden aunt.

JANE GOODALL, 'Gigi', *Through a Window: 30 Years with the Chimpanzees of Gombe*, 1990

The storm grew until sheet lightning spread across the western sky. The thunderhead reared up like a top-heavy monster in slow motion, tilted forward, blotting out the stars. The forest erupted in a simulation of violent life. Lightning bolts broke to the front and then closer, to the right and left, 10,000 volts dropping along an ionizing path at 800 kilometers an hour, kicking a counter-surge skyward ten times faster, back and forth in a split second, the whole perceived as a single flash and crack of sound. The wind freshened, and rain came stalking through the forest.

In the midst of chaos something to the side caught my attention. The lightning bolts were acting like strobe flashes to illuminate the wall of the rain forest. At intervals I glimpsed the storied structure: top canopy 30 meters off the ground, middle trees spread raggedly below that, and a lowermost scattering of shrubs and small trees. The forest was framed for a few moments in this theatrical setting. Its image turned surreal, projected into the unbounded wildness of the human imagination, thrown back in time 10,000 years. Somewhere close I knew spear-nosed bats flew through the tree crowns in search of fruit, palm vipers coiled in ambush in the roots of orchids, jaguars walked the river's edge; around them eight hundred species of trees stood, more than are native to all of North America; and a thousand species of butterflies, 6 percent of the entire world fauna, waited for the dawn.

About the orchids of that place we knew very little. About flies and beetles almost nothing, fungi nothing, most kinds of organisms nothing. Five thousand kinds of bacteria might be found in a pinch of soil, and about them we knew absolutely nothing. This was wilderness in the sixteenth-century sense, as it must have formed in the minds of the Portuguese explorers, its interior still largely unexplored and filled with strange, myth-engendering plants and animals. From such a place the pious naturalist would send long respectful letters to royal patrons about the wonders of the new world as testament to the glory of God. And I thought: there is still time to see this land in such a manner . . .

The storm arrived, racing from the forest's edge, turning from scattered splashing drops into sheets of water driven by gusts of wind. It forced me back to the shelter of the corrugated iron roof of the open-air living quarters, where I sat and waited with the *matèiros*. The men stripped off their clothing and walked out into the open, soaping and rinsing themselves in the torrential rain, laughing and singing. In bizarre counterpoint, leptodactylid frogs struck up a loud and monotonous honking on the forest floor close by. They were all around us. I wondered where they had been during the day. I had never encountered a single one while sifting through the vegetation and rotting debris on sunny days, in habitats they are supposed to prefer.

Farther out, a kilometer or two away, a troop of red howler monkeys chimed in, their chorus one of the strangest sounds to be heard in all of nature, as enthralling in its way as the songs of humpback whales. A male opened with an accelerating series of deep grunts expanding into prolonged roars and was then joined by the higher-pitched calls of the females. This far away, filtered through dense foliage, the full chorus was machine-like: deep, droning, metallic.

Such raintime calls are usually territorial advertisements, the means by which the animals space themselves out and control enough land to forage and breed. For me they were a celebration of the forest's vitality: *Rejoice! The powers of nature are within our compass, the storm is part of our biology!*

For that is the way of the nonhuman world. The greatest powers of the physical environment slam into the resilient forces of life, and nothing much happens. For a very long time, 150 million years, the species within the rain forest evolved to absorb precisely this form and magnitude of violence. They encoded the predictable occurrence of nature's storms in the letters of their genes. Animals and plants have come to use heavy rains and floods routinely to time episodes in their life cycle. They threaten rivals, mate, hunt prey, lay eggs in new water pools, and dig shelters in the rain-softened earth.

On a larger scale, the storms drive change in the whole structure of the forest. The natural dynamism raises the diversity of life by means of local destruction and regeneration.

Somewhere a large horizontal tree limb is weak and vulnerable, covered by a dense garden of orchids, bromeliads, and other kinds of plants that grow on trees. The rain fills up the cavities enclosed by the axil sheaths of the epiphytes and soaks the humus and clotted dust around their roots. After years of growth the weight has become nearly unsupportable. A gust of wind whips through or lightning strikes the tree trunk, and the limb breaks and plummets down, clearing a path to the ground. Elsewhere the crown of a giant tree emergent above the rest catches the wind and the tree sways above the rain-soaked soil. The shallow roots cannot hold, and the entire tree keels over. Its trunk and canopy arc downward like a blunt ax, shearing through smaller trees and burying understory bushes and herbs. Thick lianas coiled through the limbs are pulled along. Those that stretch to other trees act as hawsers to drag down still more vegetation. The massive root system heaves up to create an instant mound of bare soil. At yet another site, close to the river's edge, the rising water cuts under an overhanging bank to the critical level of gravity, and a 20-meter front collapses. Behind it a small section of forest floor slides down, toppling trees and burying low vegetation.

Such events of minor violence open gaps in the forest. The sky clears again and sunlight floods the ground. The surface temperature rises and the humidity falls. The soil and ground litter dries out and warms up still more, creating a new environment for animals, fungi, and microorganisms of a different kind from those in the dark forest interior. In the following months pioneer plant species take seed. They are very different from the young shade-loving saplings and understory shrubs of the prevailing old-stand forest. Fast-growing, small in stature, and short-lived, they form a single canopy that matures far below the upper crowns of the older trees all around. Their tissue is soft and vulnerable to herbivores. The palmate-leaved trees of the genus *Cecropia*, one of the gap-filling specialists of Central and South America, harbor vicious ants in hollow internodes of the trunk. These insects, bearing the appropriate scientific name *Azteca*, live in symbiosis with their hosts, protecting them from all predators except sloths and a few other herbivores specialized to feed on

Cecropia. The symbionts live among new assemblages of species not found in the mature forest.

All around the second-growth vegetation, the fallen trees and branches rot and crumble, offering hiding places and food to a vast array of basidiomycete fungi, slime molds, ponerine ants, scolytid beetles, bark lice, earwigs, embiopteran webspinners, zorapterans, entomobryomorph springtails, japygid diplurans, schizomid arachnids, pseudoscorpions, real scorpions, and other forms that live mostly or exclusively in this habitat. They add thousands of species to the diversity of the primary forest.

Climb into the tangle of fallen vegetation, tear away pieces of rotting bark, roll over logs, and you will see these creatures teeming everywhere. As the pioneer vegetation grows denser, the deepening shade and higher humidity again favor old-forest species, and their saplings sprout and grow. Within a hundred years the gap specialists will be phased out by competition for light, and the tall storied forest will close completely over . . .

Life is too well adapted in such places, out to the edge of the physical envelope where biochemistry falters, and too diverse to be broken by storms and other ordinary vagaries of nature. But diversity, the property that makes resilience possible, is vulnerable to blows that are greater than natural perturbations. It can be eroded away fragment by fragment, and irreversibly so if the abnormal stress is unrelieved. This vulnerability stems from life's composition as swarms of species of limited geographical distribution. Every habitat, from Brazilian rain forest to Antarctic bay to thermal vent, harbors a unique combination of plants and animals. Each kind of plant and animal living there is linked in the food web to only a small part of the other species. Eliminate one species, and another increases in number to take its place. Eliminate a great many species, and the local ecosystem starts to decay visibly. Productivity drops as the channels of the nutrient cycles are clogged. More of the biomass is sequestered in the form of dead vegetation and slowly metabolizing, oxygen-starved mud, or is simply washed away. Less competent pollinators take over as the best-adapted bees, moths, birds, bats, and other specialists drop out. Fewer seeds fall, fewer seedlings sprout. Herbivores decline, and their predators die away in close concert.

In an eroding ecosystem life goes on, and it may look superficially the same. There are always species able to recolonize the impoverished area and exploit the stagnant resources, however clumsily accomplished. Given enough time, a new combination of species—a reconstituted fauna and flora—will reinvest the habitat in a way that transports energy and materials somewhat more efficiently. The atmosphere they generate and the composition of the soil they enrich will resemble those found in comparable habitats in other parts of the world, since the species are adapted to penetrate and reinvigorate just such degenerate systems. They do so because they gain more energy and materials and leave more offspring. But the restorative power of the fauna and flora of the world as a whole depends on the existence of enough species to play that special role. They too can slide into the red zone of endangered species.

Biological diversity—'biodiversity' in the new parlance—is the key to the maintenance of the world as we know it. Life in a local site struck down by a passing storm springs back quickly because enough diversity still exists. Opportunistic species evolved for just such an occasion rush in to fill the spaces. They entrain the succession that circles back to something resembling the original state of the environment.

This is the assembly of life that took a billion years to evolve. It has eaten the storms—folded them into its genes—and created the world that created us. It holds the world steady. When I rose at dawn the next morning, Fazenda Dimona had not changed in any obvious way from the day before. The same high trees stood like a fortress along the forest's edge; the same profusion of birds and insects foraged through the canopy and understory in precise individual timetables. All this seemed timeless, immutable, and its very strength posed the question: how much force does it take to break the crucible of evolution?

EDWARD O. WILSON, 'Storm over the Amazon', *The Diversity of Life*, 1992

Fellow Creatures

*T*HE *view of planet Earth from space, 'the only exuberant thing in this part of the cosmos . . . [with] the organized, self-contained look of a live creature' as Lewis Thomas graphically described it, has become one of the symbols of a growing holistic mood in nature writing. Writers committed to the concept of 'spaceship earth' see no future for a human dominion over nature, and maybe not even for a human stewardship. Both ethically and ecologically we are part of an indivisible creation, and we must live as equal partners with our fellow creatures or not at all.*

This is an idea with ancient roots, despite its new notes of poignancy and urgency, and contains within it many of the recurrent themes of nature writing: the shared cycles of the seasons, the common awareness of territory, and the irreducible, inexplicable fringe of wildness—Annie Dillard's 'fluted fringes'—seemingly possessed by all living forms. Perhaps the authors who explore these ideas could be called neo-Romantics. Yet they also hark back much further, to the medieval naturalists who saw nature as a system of symbols and portents. The achievement of modern, philosophical biologists like Aldo Leopold, Loren Eiseley, and Lewis Thomas, has been to show how this ancestral habit is rooted not just in the kindredness of life, but in something perennially green and inquisitive in the human imagination.

Viewed from the distance of the moon, the astonishing thing about the earth, catching the breath, is that it is alive. The photographs show the dry, pounded surface of the moon in the foreground, dead as an old bone. Aloft, floating free beneath the moist, gleaming membrane of bright blue sky, is the rising earth, the only exuberant thing in this part of the cosmos. If you could look long enough, you would see the swirling of the great drifts of white cloud, covering and uncovering the half-hidden masses of land. If you had been looking for a very long, geologic time, you could have seen the continents themselves in motion, drifting apart on their crustal plates, held afloat by the fire beneath. It has the organized, self-contained look of a live creature, full of information, marvelously skilled in handling the sun.

LEWIS THOMAS, The Lives of a Cell, 1974

Certainly as far as wild animals and all the other inhabitants of the wide-open spaces are concerned, including birds, I have reached one unshakable and funereal conclusion: we have brought them nothing but despair. And for this reason I no longer wish to see the creatures I love with so strong an attachment in the enclosures that have now replaced their cages. I shall live henceforth on the memories I still have of them. I shall read about what men call their crimes: a tiger has torn a piece off his tamer; a lion, in love with his despotic and booted mistress, has killed the handsome boy who was his successful rival; a bear, maddened at being more narrowly confined in his cage than Cardinal Balue in his, has torn his keeper to pieces . . . I shall dream, far from these wild creatures, that we could do without them, that we could leave them to live where they were born. We should forget their true shapes then, and our imagination would flourish again. Our great-nephews would once more invent an indestructible fauna, which they would describe as they beheld it in their dreams, with dazzling intrepidity, as our forefathers used to do. I have in my possession a few big pages torn from an old natural-history book, and their colours are vivid, if not exact. On one of these pages there glows a fruit in the shape of a heart, reproduced life-size, which is to say, slightly larger than an ox heart. It seems to have a very porous skin, and from each pore there springs a large hair. A caption at the foot of the page informs us that the name of this fruit is the *lickie*, and that it grows in abundance on certain trees, thirty feet high, which are providentially distributed here and there across desert regions made desolate by the prevailing hunger and thirst. 'A true gift of providence for the traveller, the lickie has the taste and consistency of veal, which makes it an ideal family dish . . .' After a single perusal of those lines, are you not perfectly delighted and dazzled? And so we should be always if we were to meditate upon documents describing distant fauna that originated, not in the austere file indices of explorers, but in the exaltation of an artist. If some commandment of an all-powerful apostle could contain once more within the limits of their jungles, their savannas, and their polar regions all those beasts that now languish here with us, then I would willingly pledge myself to describe the animals thus placed beyond our reach, and to provide you with matter for your dreams. 'The Pempek, an animal already known in Ancient Times, haunts the solitudes of the Matto Grosso. It has big, flat feet which produce a musical sound as it walks, and a trunk with which it sucks in butterflies on the wing. Its mane is very thick, and it always runs away from the colour blue . . .'

The reality is otherwise, and we are no longer free even to remain ignorant of how a boa constrictor chokes a gazelle, or how a panther, deliberately starved, rips open the throat of a goat which—since the combat must be spiced somehow and the cinema has no use for passive victims—has a kid to defend. It is high time I said good-bye to reality, to such supposedly ferocious animals and such indubitably guilty men, to motionless birds standing upright on talons embedded in dung, to kangaroos yielding little by little to paralysis, and to lion cubs crippled by rickets. Where shall I find my place of solitude? There is no beautiful human face, no snowy fur, no azure pinions that can enchant me if they are branded with the intolerable and parallel shadows cast by bars . . .

> COLETTE, 'Bitterness', *En Pays Connu*, 1950, tr. Derek Coltman; from *Earthly Paradise*, ed. Robert Phelps, 1966

The Maures are my favourite mountains, a range of old rounded mammalian granite which rise three thousand feet above the coast of Provence. In summer they are covered by dark forests of cork and pine, with paler interludes on the northern slopes of bright splay-trunked chestnut, and an undergrowth of arbutus and bracken. There is always water in the Maures, and the mountains are green throughout the summer, never baked like the limestone, or like the Southern Alps a slagheap of gritty oyster-shell. They swim in a golden light in which the radiant ebony green of their vegetation stands out against the sky, a region hardly inhabited, yet friendly as those dazzling landscapes of Claude and Poussin, in which shepherds and sailors from antique ships meander under incongruous elms. Harmonies of light and colour, drip of water over fern; they inculcate in those who stay long in the Midi, and whose brains are addled by iodine, a habit of moralizing, a brooding about causes. What makes men divide up into nations and go to war? Why do they live in cities? And what is the true relationship between Nature and Man?

The beaches of the Maures are of white sand, wide, with a ribbon of umbrella-pines, below which juicy mesembrianthemum and dry flowers of the sand stretch to within a yard of the sea. Lying there amid the pacific blues and greens one shuts the eyes and opens them on the white surface: the vague blurred philosophizing continues. Animism, pantheism, images of the earth soaring through space with the swerve of a ping-pong ball circulate in the head; the woolly brain meddles with ethics. No more power, no aggression, no intolerance. All must be free. Then whizz! A disturbance. Under the eye the soil is pitted into a conical depression, about the size of a candle extinguisher, down whose walls the sand trickles gently, moved by a suspicion of wind. Whizz, and a clot is hurled to the top again, the bottom of the funnel cleared, in disobedience to the natural law! As the funnel silts up it is cleared by another whirr, and there appears, at the nadir of the cone, a brown pair of curved earwig horns, antlers of a giant earwig that churn the sand upwards like a steam shovel.

Now an ant is traversing the dangerous *arête*. He sidles, slithers, and goes fumbling down the Wall of Death to the waiting chopper. Snap! He struggles up, mounting the steep banking grain by grain as it shelves beneath him, till a new eruption is engineered by his waiting enemy. Sand belches out, the avalanche engulfs him, the horny sickles contract and disappear with their beady victim under the whiteness. Mystery, frustration, tragedy, death are then at large in this peaceful wilderness! Can the aggressive instinct be analysed out of those clippers? Or its lethal headpiece be removed by a more equitable distribution of raw materials? The funnels, I observe, are all round me. The sand is pockmarked with these geometrical death-traps, engineering triumphs of insect art. And this horsefly might be used for an experiment. I shove it downwards. The Claws seize on a wing, and the struggle is on. The fight proceeds like an atrocity of chemical warfare. The great fly threshes the soil with its wings, it buzzes and drones while the sand heaves round its propellers and the facets of its giant projectors glitter with light. But the clippers do not relax, and disappear tugging the fly beneath the surface. The threshing continues, a faint buzzing comes from the invisible horsefly, and its undercarriage appears, with legs waving. Will it take off? The wings of the insect bomber pound the air, the fly starts forward and upwards, and hauls after it—O fiend, embodiment of evil! A creature whose clippers are joined to a musclebound thorax and a vile yellow armour-plated body, squat and powerful, with a beetle set of legs to manoeuvre this engine of destruction. The Tank with a Mind now scuttles backwards in reverse, the stern, then the legs disappear, then the jaws which drag its prey. Legs beat the ground. A fainter wheeze and whirr, no hope now, the last wing-tip vanished, the air colder, the pines greener, the cone empty except for the trickle, the sifting and silting down the funnel of the grains of pearl-coloured sand.

Nature arranged this; bestowed on the Ant-Lion its dredging skill and its cannon-ball service. How can it tell, buried except for the striking choppers, that the pebble which rolls down has to be volleyed out of the death-trap, while the approaching ant must be collected by gentle eruptions, dismayed by a perpetual sandy shower? And, answer as usual, we do not know.

CYRIL CONNOLLY, 'The Ant-Lion', *The Condemned Playground*, 1945

On my way home I picked up a dead bird. Having just gazed into the cradle of life, I felt a desire to take home the dead body and watch with like attention the activities of this poor discarded garment that was now the cloth of death. I put it in a basin and left it in a shed. Returning after a week, I found it had come to life again. It was breathing heavily. Its tongue popped in and out of its beak, its eyes flashed, and it made a grinding noise. This surprised me; but I then saw that the tongue was really a white worm, the flashing eye a white worm, while the body heaved owing to the squirming activity of the pack of worms inside the corpse.

To find the explanation of this we need go no farther than the female bluebottle on the point of laying her eggs. She prefers to lay them in meat, in a hole in the meat, which will serve as cradle and as food. For this purpose she finds nothing so good as dead birds. The procedure is as follows. She approaches the corpse and makes straight for the beak. If it is tightly closed she will go to the eye-sockets, but if it is open she thrusts her egg-conducting tube, her oviduct, into the hole and proceeds to lay her eggs, an operation which, allowing for rests from labour, may take two hours—after which she goes away and dies. The bird's beak has now been packed pretty full, the tongue and throat being white with layers of eggs. Here they remain for two days, after which time they are transformed into maggots who then descend down the throat of the bird.

I made my examination several days after they had gone down there and had been composing themselves while decomposing the bird. Indeed they had so completely taken possession that the whole body heaved about, and some of these white, squirming maggots, like small spaghetti, had returned to the throat and also entered the eye-sockets. Already the body had lost much of its weight, for death is heavy and life is light.

I opened the flesh a bit more so that I might observe the main work of reconstruction. I gazed down at the living tubes as they squirmed and twisted and turned and turned them at their task, building up new life in the abominable ferment of corruption. The bluebottle is necessary. The bluebottle is good. All things in Nature have a meaning and a purpose. All are necessary. All are right. If it were not so, if any one thing were wrong, then nothing could be right; if a single error marred the scheme then we could count on nothing, all would be lost, we could hope for nothing, there would be nothing for it, as Edward Carpenter said, 'but to fold our hands and be damned everlastingly'. But since it is not so we can afford to face the facts. It is expedient, on occasion, to gaze down into the pit as well as up towards heaven, to look at the roots of Nature as well as at her flags, regarding the burden of the beginning and the dereliction of the end alike without flinching, so that from time to time the seeing eye, the accepting mind, may receive the vision of what some men call beauty and others truth.

JOHN STEWART COLLIS, 'The Scavengers of Corruption',
Down to Earth, 1947

The love of Nature is deep in England. And I think that what is behind this love is the instinct that Nature has a secret for us, and answers our questions. Take that foxglove over there—for we have now reached August in this chronicle. It stands singly where there had been such a wonderful display of bluebells that it then looked as if a section of the sky had been established upon earth (though not really the same colour at all!). That foxglove with its series of petal-made thimbles held up for sale to the bees, puts me at ease upon the subject of—progress. It is quite obvious that the foxglove cannot be *improved*. There is no progressing beyond that point for that particular Appearance. There is no room for improvement in the bluebell nor in any of the other exhibits. The fact is we get perfection in this form and in that form. Hence Shakespeare's 'ripeness is all', and Tennyson's 'God fulfils Himself in many ways', and Whitman's 'there can never be any more perfection than there is now', and Heraclitus' 'Life is a Fountain of Fire, an ever-living Flame, kindled in due measure, and in like measure extinguished'. Evolution is not something going up and up and up—but a series of perfect Forms. The goal of each Form is the fulfilment of its own unique perfection. There is no point in our gazing raptly into the future for paradise if it is at our feet.

But this is not true of Man, you say. That is the paradox. In a perfect world he is imperfect. But then he has attained a new thing of his own—consciousness. Complete consciousness will be his ripeness, his perfection. That will probably take time, say several million years. But why worry? there might be five million years after that of perfect humanity. Meanwhile our foxglove can keep us sane at least about subjects such as beauty and art. There is no steady evolutionary 'progress' in these things, only different expressions. Just as there will never be a better foxglove so there will never be a better Shakespeare.

JOHN STEWART COLLIS, 'The Feeling Intellect', *Down to Earth*, 1947

Before the swallow, before the daffodil, and not much later than the snowdrop, the common toad salutes the coming of spring after his own fashion, which is to emerge from a hole in the ground, where he has lain buried since the previous autumn, and crawl as rapidly as possible towards the nearest suitable patch of water. Something—some kind of shudder in the earth, or perhaps merely a rise of a few degrees in the temperature—has told him that it is time to wake up: though a few toads appear to sleep the clock round and miss out a year from time to time—at any rate, I have more than once dug them up, alive and apparently well, in the middle of the summer.

At this period, after his long fast, the toad has a very spiritual look, like a strict Anglo-Catholic towards the end of Lent. His movements are languid but purposeful, his body is shrunken, and by contrast his eyes look abnormally large. This allows one to notice, what one might not at another time, that a toad has about the most beautiful eye of any living creature. It is like gold, or more exactly it is like the golden-coloured semi-precious stone which one sometimes sees in signet-rings, and which I think is called a chrysoberyl.

For a few days after getting into the water the toad concentrates on building up his strength by eating small insects. Presently he has swollen to his normal size again, and then he goes through a phase of intense sexiness. All he knows, at least if he is a male toad, is that he wants to get his arms round something, and if you offer him a stick, or even your finger, he will cling to it with surprising strength and take a long time to discover that it is not a female toad. Frequently one comes upon shapeless masses of ten or twenty toads rolling over and over in the water, one clinging to another without distinction of sex. By degrees, however, they sort themselves out into couples, with the male duly sitting on the female's back. You can now distinguish males from females, because the male is smaller, darker and sits on top, with his arms tightly clasped round the female's neck. After a day or two the spawn is laid in long strings which wind themselves in and out of the reeds and soon become invisible. A few more weeks, and the water is alive with masses of tiny tadpoles which rapidly grow larger, sprout hind-legs, then forelegs, then shed their tails: and finally, about the middle of the summer, the new generation of toads, smaller than one's thumb-nail but perfect in every particular, crawl out of the water to begin the game anew.

I mention the spawning of the toads because it is one of the phenomena of spring which most deeply appeal to me, and because the toad, unlike the skylark and the primrose, has never had much of a boost from the poets. But I am aware that many people do not like reptiles or amphibians, and I am not suggesting that in order to enjoy the spring you have to take an interest in toads. There are also the crocus, the missel-thrush, the cuckoo, the blackthorn, etc. The point is that the pleasures of spring are available to everybody, and cost nothing. Even in the most sordid street the coming of spring will register itself by some sign or other, if it is only a brighter blue between the chimney pots or the vivid green of an elder sprouting on a blitzed site. Indeed it is remarkable how Nature goes on existing unofficially, as it were, in the very heart of London. I have seen a kestrel flying over the Deptford gasworks, and I have heard a first-rate performance by a blackbird in the Euston Road. There must be some hundreds of thousands, if not millions, of birds living inside the four-mile radius, and it is rather a pleasing thought that none of them pays a halfpenny of rent.

As for spring, not even the narrow and gloomy streets round the Bank of England are quite able to exclude it. It comes seeping in everywhere, like one of those new poison gases which pass through all filters. The spring is commonly referred to as 'a miracle', and during the past five or six years this worn-out figure of speech has taken on a new lease of life. After the sort of winters we have had to endure recently, the spring does seem miraculous, because it has become gradually harder and harder to believe that it is actually going to happen. Every February since 1940 I have found myself thinking that this time winter is going to be permanent. But Persephone, like the toads, always rises from the dead at about the same moment. Suddenly, towards the end of March, the miracle happens and the decaying slum in which I live is transfigured. Down in the square the sooty privets have turned bright green, the leaves are thickening on the chestnut trees, the daffodils are out, the wallflowers are budding, the policeman's tunic looks positively a pleasant shade of blue, the fishmonger greets his customers with a smile, and even the sparrows are quite a different colour, having felt the balminess of the air and nerved themselves to take a bath, their first since last September.

Is it wicked to take a pleasure in spring and other seasonal changes? To put it more precisely, is it politically reprehensible, while we are all groaning, or at any rate ought to be groaning, under the shackles of the capitalist system, to point out that life is frequently more worth living because of a blackbird's song, a yellow elm tree in October, or some other natural phenomenon which does not cost money and does not have what the editors of left-wing newspapers call a class angle? There is no doubt that many people think so. I know by experience that a favourable reference to 'Nature' in one of my articles is liable to bring me abusive letters, and though the key-word in these letters is usually 'sentimental', two ideas seem to be mixed up in them. One is that any pleasure in the actual process of life encourages a sort of political quietism. People, so the thought runs, ought to be discontented, and it is our job to multiply our wants and not simply to increase our enjoyment of the things we have already. The other idea is that this is the age of machines and that to dislike the machine, or even to want to limit its domination, is backward-looking, reactionary and slightly ridiculous. This is often backed up by the statement that a love of Nature is a foible of urbanised people who have no notion what Nature is really like. Those who really have to deal with the soil, so it is argued, do not love the soil, and do not take the faintest interest in birds or flowers, except from a strictly utilitarian point of view. To love the country one must live in the town, merely taking an occasional week-end ramble at the warmer times of year.

This last idea is demonstrably false. Medieval literature, for instance, including the popular ballads, is full of an almost Georgian enthusiasm for Nature, and the art of agricultural peoples such as the Chinese and Japanese centres always round trees, birds, flowers, rivers, mountains. The other idea seems to me to be wrong in a subtler way. Certainly we ought to be discontented, we ought not simply to find out ways of making the best of a bad job, and yet if we kill all pleasure in the actual process of life, what sort of future are we preparing for ourselves? If a man cannot enjoy the return of spring, why should he be happy in a labour-saving Utopia? What will he do with the leisure that the machine will give him? I have always suspected that if our economic and political problems are ever really solved, life will become simpler instead of more complex, and that the sort of pleasure one gets from finding the first primrose will loom larger than the sort of pleasure one gets from eating an ice to the tune of a Wurlitzer. I think that by retaining one's childhood love of such things as trees, fishes, butterflies and—to return to my first instance—toads, one makes a peaceful and decent future a little more probable, and that by preaching the doctrine that nothing is to be admired except steel and concrete, one merely makes it a little surer that human beings will have no outlet for their surplus energy except in hatred and leader worship.

At any rate, spring is here, even in London N.1, and they can't stop you enjoying it. This is a satisfying reflection. How many a time have I stood watching the toads mating, or a pair of hares having a boxing match in the young corn, and thought of all the important persons who would stop me enjoying this if they could. But luckily they can't. So long as you are not actually ill, hungry, frightened or immured in a prison or a holiday camp, spring is still spring. The atom bombs are piling up in the factories, the police are prowling through the cities, the lies are streaming from the loudspeakers, but the earth is still going round the sun, and neither the dictators nor the bureaucrats, deeply as they disapprove of the process, are able to prevent it.

GEORGE ORWELL, 'Some Thoughts on the Common Toad', *Tribune*, 12 April 1946: from *The Collected Essays of George Orwell*, vol. iv, 1968

Welcome to our village and our woods. I welcome you first to our woods, because they are the oldest. Before there were men in Abinger, there were trees. Thousands of years before the Britons came, the ash grew at High Ashes and the holly at Holmwood and the oak at Blindoak Gate; there were yew and juniper and box on the downs before ever the Pilgrims came along the Pilgrims' way. They greet you, and our village greets you.

What shall we show you? History? Yes, but the history of a village lost in the woods. Do not expect great deeds and grand people here. Lord and Ladies, warriors and priests will pass, but this is not their home, they will pass like the leaves in autumn but the trees remain. The trees built our first houses and our first church, they roof our church today, they are with us from the cradle to the grave.

E. M. FORSTER, prologue to a 'Pageant of 12 Woods' held in Abinger, n.d. (quoted in P. N. Furbank, *E. M. Forster: A Life*, 1978)

When I was six or seven, being brought up, at that time, by my grandmother—my mother having died a year or so before—we used to walk the mile or two to various spots on the shore, carrying bucket and spade, baskets and bags, and all that was necessary for the picnic. A jugful of boiling water, for making tea, could be obtained, price one penny, from one of the houses near by. It would never have occurred to my grandmother that you could drink anything but tea.

Yet I did not think of myself as living anywhere but in a town. I was in my own eyes, essentially a town boy, like the boys of Barrow-in-Furness, or, if it came to that, of London. The country was where the farmers lived. Admittedly, the farmers' sons came to our school, and one of them shared a desk with me. I remember that once we were told to draw a turnip from memory. Most of us, including me, drew a football with an aspidistra sprouting from the top, but I noticed that my neighbour had given *his* football a topping of crinkly leaves like the fronds of an arthritic fern. I had no great regard for his artistic abilities but it did strike me that he might know what a turnip looked like, so I rubbed out my leaves and copied his. Later in the lesson I was commended by the teacher for my powers of observation!

It was round about that time that certain plants must have impressed themselves upon my consciousness. They mostly belonged to high summer and August, and, even today, the sudden sight of one of them can bring back the brightness, the smell, the exhilaration of school holidays. For the most part I did not know their names, or, if I did, I had learnt them from cigarette cards—ragwort, tufted vetch, silverweed, and the harebell, which, like the Scots, we called 'bluebell'. With these I remember also the stalks of thistle growing in a parched brown pasture, with a doily of green around the base of the plant where the cattle had not been able to graze. The image is as clear as if I had seen it yesterday. In fact, as I write in August, I *did* see it yesterday.

A year or two later, my father remarried, and my stepmother was what was then known as a 'great walker'—though not, of course, a 'hiker', for I don't think the term had yet been invented. I now began to learn more of the country within two or three miles of Millom. On bank holidays we sometimes took the train from Foxfield to Coniston; on Ascension Day we climbed Black Combe, the great hill above Millom which can be seen from Scotland, Wales and the Isle of Man. Above all, every Sunday morning, my father and I went for a walk through the Hodbarrow Mines.

Hodbarrow, in its time, had been one of the most famous iron mines in the world, working a deposit of haematite which, until the end of the last century, was the richest ever exploited. By the 1920s, it was already in decline, though it continued to produce ore for another forty years. There were twenty or thirty abandoned shafts and about half a dozen still working—all spread over a huge acreage of limestone outcrop at the mouth of the Duddon. In the early part of this century, a mile-and-a-quarter-long sea wall had been thrown out, like the arch of a bow, enclosing a new parish of shingle, sand and sump. Gradually, as it was undermined, the land subsided, turning the enclosed hollow into a great sunken basin. It was out of bounds in my boyhood, but, since then, I have walked over every yard of it, searching for the sand and limestone flowers which hurried to colonise the new habitat in the sixty years between the high tides of yesterday and the permitted flooding of today. I could stand at the lowest point in Cumbria, seventy feet below sea level, and look up to Scafell Pike, the highest point in England.

Mining is really a rural industry—the harvesting of a crop which, unfortunately, does not renew itself. An iron ore mine, in particular, tends to look like rough, untidy farming on humpy and intractable land. For one thing, unlike coal, iron ore is not found in seams, spread flatly, like jam in a sandwich, between the rock strata. It lies in solid globules or blobs, which may contain many thousands of tons of metal and yet be encompassed within a surface area no bigger than a tennis court. In the old days, prospecting was such a hit-and-miss affair that they could drop boreholes all round a deposit, like a knife-thrower just missing his accomplice, and fail to probe to the spot where the iron lay.

All this meant that, in a large royalty like Hodbarrow, between all the pit-heads, the working and worked-out shafts, there was a scattering and jigsaw of broken ground where nature, from the beginning, had been fighting back. Again, unlike coal, iron ore mines are not dirty. The iron dust is red; old railway sleepers and wagons are stained a dark cochineal. The rubble tips and spoil heaps—*never* call them 'slagbanks'!—are like dark red tumuli, greening and weathering with grass and horsetail. At Hodbarrow, the rock—a beautiful white limestone, tinged with pink—could be seen breaking through the surface, quarried and blasted into cliffs and chasms and potholes. Little mineral lines ran up and down the hummocky ground like a roller-coaster at a fair. Where the soil had lain undisturbed for long, there were rank thickets of hawthorn and blackthorn, gorse and broom. On the old foreshore, reclaimed by the sea-wall, there was heather and dwarf willow and sea-holly. There were hares, foxes and badgers. The now rare natterjack toad bred by the thousand. In that part of the hollow where the sand was earthiest, the managers had grown a crop of mangels to give cover to the partridge which they shot once a year. The men, too, had their wildlife interests. A cuckoo's egg, laid in the nest of some small bird, was watched and guarded till the young cuckoo was fledged and flew away. My step-grandfather, who was in charge of one of the surface engines that supplied power below ground, would come home, in summer, with a sprig of centaury in his buttonhole. He died in 1931, but the tansy which he planted outside his engine-house still grows on the abandoned site.

Of course, when I was a boy, I did not give a thought to this now seemingly remarkable closeness of wild nature to industry, of country to town. I don't think I ever thought of the mines as being country at all. I saw them as a place where men worked. For, even on a Sunday, there would still be perhaps a single loco-motive, fat and puffy as a pack mule, dragging a few wagons up a pit-bank, and now and again the pit-head wheels would start whirring to raise a couple of bogies or to lower two of the bosses to inspect the underground levels. That a whole flora of bushes and shrubs and weeds and mosses should be growing, often within a few yards of the mine workings, was not a matter for surprise.

When I grew older, this closeness between wild nature and everyday life took on a new meaning. Millom, I saw, came out of the rock, grew on it like a tree and was entirely dependent on it. Physically, indeed, the town is—or was, before the new hous-ing of the 1930s—as much a product of Cumbrian rock as any of the dale villages. The older houses, the schools, the chapels, the town hall are all built of stone from two nearby quarries; they are roofed with slates from the quarries of Kirkby-in-Furness, on the other side of the Duddon. The town church is of sandstone from St Bees, only twenty miles up the coast; the railway station of Eskdale granite, from still nearer home. The limestone of Hodbarrow, oddly enough, does not show itself in the town build-ings at all.

Yet it was from this limestone that the town came into being, and it was here that a kind of topographical emblem began to shape itself in my mind. Everywhere, and at all times, men are dependent on the physical world they live on. Here, in Millom of the mining days, you could see that pattern spread before you,

you could read the history of the Industrial Revolution inscribed on the landscape. The ore came out of the rock; was shunted across to the ironworks; was smelted into pig-iron, sold and turned into money. The whole life of the town came from that first product of the rock. The ore was rightly called 'haematite' for it was the life-blood of the whole community. It is the same everywhere. Even if we feed on vitamin tablets and live in plastic houses, our life will still come from rock, wind, air and sun, though it is hard to remember this in a city. At Hodbarrow, you can't forget it.

NORMAN NICHOLSON, 'Ten-Yard Panorama', *Second Nature*, ed. R. Mabey, 1984

A couple of years after the end of the war I got my first semi-job on Fleet Street—as supernumerary reporter on a Sunday newspaper. I was a sort of journalistic char brought in to clear up the week's news offscourings, too menial for the splendid by-liners and feature writers who strode past my little ghetto of desks to interview a Minister or lunch with a film star . . .

But after only a few weeks of rewriting news-in-brief paragraphs and vainly ringing the door bells of criminals' wives, the greatest pleasure of my dismal duty day came to be the morning walk from High Holborn down Shoe Lane. During that spring I always looked for the kestrels. Then, stretching from the Farringdon Road to Fetter Lane was a vast moraine of rubble where in the blitz a stick of bombs had swiped down an entire district. Already, in the basement and alleys and dim print shops laid open by the Luftwaffe to the sun—in that riding scythed through the brick forest—wild flowers had unfurled like bunting across the rawness. Rose-bay willow-herb spread in purple surf over the brick piles, and bracken fern too; and upon the cracked flags of what had been cellars were bright yellow mats of ragwort and coltsfoot. Insects and house sparrows and mice rummaged in the new wilderness, and so there was food for the hawks.

On the pavement by the Two Brewers pub I watched them—flickering arrowheads—quartering their territory on chestnut-red, slender wings and hovering at bus-roof height. They obviously had an eyrie somewhere in the gutted buildings at the edge of the blast area, for late in May I saw them carrying prey over there to where there would be young on a ledge among the ruins. They filled me with an elegiac delight, those wild kestrels: like the ones whose eggs I took and whose eyasses I had reared for novice falconry in the pre-war parkland I had ranged, and which I now saw through the eyes of a captive bird in a glass cage, living on stale newsdesk crumbs.

KENNETH ALLSOP, 'The Wildlife of London', *Sunday Times Magazine*, 23 March 1969

On the top of a stepladder, I made one more observation upon life. It was cold that autumn evening, and, standing under a suburban street light in a spate of leaves and beginning snow, I was suddenly conscious of some huge and hairy shadows dancing over the pavement. They seemed attached to an odd, globular shape that was magnified above me. There was no mistaking it. I was standing under the shadow of an orb-weaving spider. Gigantically projected against the street, she was about her spinning when everything was going underground. Even her cables were magnified upon the sidewalk and already I was half-entangled in their shadows.

'Good Lord,' I thought, 'she has found herself a kind of minor sun and is going to upset the course of nature.'

I procured a ladder from my yard and climbed up to inspect the situation. There she was, the universe running down around her, warmly arranged among her guy ropes attached to the lamp supports—a great black and yellow embodiment of the life force, not giving up to either frost or stepladders. She ignored me and went on tightening and improving her web.

I stood over her on the ladder, a faint snow touching my cheeks, and surveyed her universe. There were a couple of iridescent green beetle cases turning slowly on a loose strand of web, a fragment of luminescent eye from a moth's wing and a large indeterminable object, perhaps a cicada, that had struggled and been wrapped in silk. There were also little bits and slivers, little red and blue flashes from the scales of anonymous wings that had crashed there.

Some days, I thought, they will be dull and gray and the shine will be out of them; then the dew will polish them again and drops hang on the silk until everything is gleaming and turning in the light. It is like a mind, really, where everything changes but remains, and in the end you have these eaten-out bits of experience like beetle wings.

I stood over her a moment longer, comprehending somewhat reluctantly that her adventure against the great blind forces of winter, her seizure of this warming globe of light, would come to nothing and was hopeless. Nevertheless it brought the birds back into my mind, and that faraway song which had traveled with growing strength around a forest clearing years ago—a kind of heroism, a world where even a spider refuses to lie down and die if a rope can still be spun on to a star. Maybe man himself will fight like this in the end, I thought, slowly realizing that the web and its threatening yellow occupant had been added to some luminous store of experience, shining for a moment in the fog-bound reaches of my brain.

The mind, it came to me as I slowly descended the ladder, is a very remarkable thing; it has gotten itself a kind of courage by looking at a spider in a street lamp. Here was something that ought to be passed on to those who will fight our final freezing battle with the void. I thought of setting it down carefully as a message to the future: *In the days of the frost seek a minor sun.*

But as I hesitated, it became plain that something was wrong. The marvel was escaping—a sense of bigness beyond man's power to grasp, the essence of life in its great dealings with the universe. It was better, I decided, for the emissaries returning from the wilderness, even if they were merely descending from a stepladder, to record their marvel, not to define its meaning. In that way it would go echoing on through the minds of men, each grasping at that beyond out of which the miracles emerge, and which, once defined, ceases to satisfy the human need for symbols.

In the end I merely made a mental note: One specimen of Epeira observed building a web in a street light. Late autumn and cold for spiders. Cold for men, too. I shivered and left the lamp glowing there in my mind. The last I saw of Epeira she was hauling steadily on a cable. I stepped carefully over her shadow as I walked away.

Loren Eiseley, 'The Judgement of the Birds', The Immense Journey, 1946

The wind picks up and blusters. Its fat underbelly scrapes the uneven ground, twisting like taffy towards me, slips up over the mountain and showers out across the Great Plains. The sea smell it carried all the way from Seattle has long since been absorbed

by pink grus—the rotting granite that spills down the slopes of
the Rockies. Somewhere over the Midwest the wind slows, tan-
gling in the hair of hardwood forests, and finally drops into the
corridors of the cities, past Manhattan's World Trade Center,
ripping free again as it crosses the Atlantic's green swell.

Spring jitterbugs inside me. Spring *is* wind, symphonic and billowing. A dark cloud pops like a blood blister over me, letting
hail down. It comes on a piece of wind that seems to have widened
the sky, comes so the birds have something to fly on.

A message reports to my brain but I can't believe my eyes.
The sheet of wind had a hole in it: an eagle just fell out of the
sky. It fell as if down the chute of a troubled airplane. Landed,
falling to one side as if a leg were broken. I was standing on the
hill overlooking the narrow valley that had been a seashore
170,000,000 years ago, whose sides had lifted like a medic's litter to catch up this eagle now.

She hops and flaps seven feet of wing and closes them down
and sways. She had come down (on purpose?) near a dead fawn
whose carcass had recently been feasted upon. When I walked
closer, all I could see of the animal was a ribcage rubbed red
with fine tissue and the decapitated head lying peacefully against
sagebrush, eyes closed.

At twenty yards the eagle opened her wings halfway and rose
up, her whole back lengthening and growing stiff. At forty feet
she looked as big as a small person. She craned her neck, first
to one side, then the other, and stared hard. She's giving me 'the
eagle eye,' I thought.

Friends who have investigated eagles' nests have literally feared
for their lives. It's not that they were in danger of being pecked
to death but, rather, grabbed. An eagle's talons are a powerful
jaw. Their grip is so strong the talons can slice down through
flesh to bone in one motion.

But I had come close only to see what was wrong, to see what
I could do. An eagle with a bum leg will starve to death. Was it
broken, bruised, or sprained? How could I get close enough to
know? I approached again. She hopped up in the air dashing the
critical distance between us with her great wings. Best to leave
her alone, I decided. My husband dragged a road-killed deer up
the mountain slope so she could eat, and I brought a bucket of
water. Then we turned towards home.

A golden eagle is not golden but black with yellow spots on

the neck and wings. Looking at her, I had wondered how feathers came to be, how their construction—the rachis, vane, and quill—is unlike anything else in nature.

Birds are glorified flying lizards. The remarkable feathers which, positioned together, are like hundreds of smaller wings, evolved from reptilian scales. Ancestral birds had thirteen pairs of cone-shaped teeth that grew in separate sockets like a snake's, rounded ribs, and bony tails. Archaeopteryx was half bird, half dinosaur who glided instead of flying; Ichthyornis was a fish-bird, a relative of the pelican; Diatryma was a giant, seven feet tall with a huge beak and wings so absurdly small they must have been useless, though later the wingbone sprouted from them. *Aquila chrysaëtos*, the modern golden eagle, has seven thousand contour feathers, no teeth, and weighs about eight pounds.

I think about the eagle. How big she was, how each time she spread her wings it was like a thought stretching between two seasons.

Back at the house I relax with a beer. At 5:03 the vernal equinox occurs. I go outside and stand in the middle of a hayfield with my eyes closed. The universe is restless but I want to feel celestial equipoise: twelve hours of daylight, twelve of dark, and the earth ramrod straight on its axis. In celebration I straighten my posture in an effort to resist the magnetic tilt back into dormancy, spiritual and emotional reticence. Far to the south I imagine the equatorial sash, now nose to nose with the sun, sizzling like a piece of bacon, then the earth slowly tilting again.

In the morning I walk up to the valley again. I glass both hillsides, back and forth through the sagebrush, but the eagle isn't there. The hindquarters of the road-killed deer have been eaten. Coyote tracks circle the carcass. Did they have eagle for dinner too?

Afternoon. I return. Far up on the opposite hill I see her, flapping and hopping to the top. When I stop, she stops and turns her head. Her neck is the plumbline on which earth revolves. Even at two hundred yards, I can feel her binocular vision zeroing in; I can feel the heat of her stare.

Later, I look through my binoculars at all sorts of things. I'm seeing the world with an eagle eye. I glass the crescent moon. How jaded I've become, taking the moon at face value only, forgetting the charcoal, shaded backside, as if it weren't there at all.

That night I dream about two moons. One is pink and spins

fast; the other is an eagle's head, farther away and spinning in the opposite direction. Slowly, both moons descend and then it is day.

At first light I clamber up the hill. Now the dead deer my husband brought is only a hoop of ribs, two forelegs, and hair. The eagle is not here or along the creek or on either hill. I go to the hill and sit. After a long time an eagle careens out from the narrow slit of the red-walled canyon whose creek drains into this valley. Surely it's the same bird. She flies by. I can hear the bonecreak and whoosh of air under her wings. She cocks her head and looks at me. I smile. What is a smile to her? Now she is not so much flying as lifting above the planet, far from me.

GRETEL EHRLICH, 'Spring', *Antaeus on Nature*, ed. Daniel Halpern, 1986

Nature is unlike art in terms of its product—what we in general know it by. The difference is that it is not only created, an external object with a history, and so belonging to a past; but also creating in the present, as we experience it. As we watch, it is so to speak rewriting, reformulating, repainting, rephotographing itself. It refuses to stay fixed and fossilized in the past, as both the scientist and the artist feel it somehow ought to; and both will generally try to impose this fossilization on it.

Verbal tenses can be very misleading here: we stick adamantly in speech to the strict protocol of actual time. Of and in the present we speak in the present, of the past in the past. But our psychological tenses can be very different. Perhaps because I am a writer (and nothing is more fictitious than the past in which the first, intensely alive and present, draft of a novel goes down on the page), I long ago noticed this in my naturalist self: that is, a disproportionately backward element in any present experience of nature, a retreat or running-back to past knowledge and experience, whether it was the definite past of personal memory or the indefinite, the imperfect, of stored 'ological' knowledge and proper scientific behaviour. This seemed to me often to cast a mysterious veil of deadness, of having already happened, over the actual and present event or phenomenon.

I had a vivid example of it only a few years ago in France, long after I thought I had grown wise to this self-imposed brainwashing. I came on my first Soldier Orchid, a species I had long

wanted to encounter, but hitherto never seen outside a book. I
fell on my knees before it in a way that all botanists will know.
I identified, to be quite certain, with Professors Clapham, Tutin
and Warburg in hand (the standard British *Flora*), I measured, I
photographed, I worked out where I was on the map, for future
reference. I was excited, very happy, one always remembers one's
'firsts' of the rarer species. Yet five minutes after my wife had
finally (other women are not the only form of adultery) torn me
away, I suffered a strange feeling. I realized I had not actually
seen the three plants in the little colony we had found. Despite
all the identifying, measuring, photographing, I had managed to
set the experience in a kind of present past, a having-looked, even
as I was temporally and physically still looking. If I had the
courage, and my wife the patience, I would have asked her to
turn and drive back, because I knew I had just fallen, in the stu-
pidest possible way, into an ancient trap. It is not necessarily too
little knowledge that causes ignorance; possessing too much, or
wanting to gain too much, can produce the same result.

There is something in the nature of nature, in its presentness,
its seeming transience, its creative ferment and hidden potential,
that corresponds very closely with the wild, or green man, in our
psyches; and it is a something that disappears as soon as it is rel-
egated to an automatic pastness, a status of merely classifiable
thing, image taken *then*. 'Thing' and 'then' attract each other. If
it is thing, it was then; if it was then, it is thing. We lack trust
in the present, this moment, this actual seeing, because our cul-
ture tells us to trust only the reported back, the publicly framed,
the edited, the thing set in the clearly artistic or the clearly scientific angle of perspective. One of the deepest lessons we have
to learn is that nature, of its nature, resists this. It waits to be
seen otherwise, in its individual presentness and from our individual presentness.

I come now near the heart of what seems to me to be the single greatest danger in the rich legacy left us by Linnaeus and the
other founding fathers of all our sciences and scientific mores
and methods—or more fairly, left us by our leaping evolutionary
ingenuity in the invention of tools. All tools, from the simplest
word to the most advanced space probe, are disturbers and
rearrangers of primordial nature and reality—are, in the dictionary definition, 'mechanical implements for working upon something.' What they have done, and I suspect in direct proportion

to our ever-increasing dependence on them, is to addict us to purpose: both to looking for purpose in everything external to us and to looking internally for purpose in everything we do—to seek explanation of the outside world by purpose, to justify our seeking by purpose. This addiction to finding a reason, a function, a quantifiable yield, has now infiltrated all aspects of our lives—and become effectively synonymous with pleasure. The modern version of hell is purposelessness.

Nature suffers particularly in this, and our indifference and hostility to it is closely connected with the fact that its only purpose appears to be being and surviving. We may think that this comprehends all animate existence, including our own; and so it must, ultimately; but we have long ceased to be content with so abstract a motive. A scientist would rightly say that all form and behaviour in nature is highly purposive, or strictly designed for the end of survival—specific or genetic, according to theory. But most of this functional purpose is hidden to the non-scientist, indecipherable; and the immense variety of nature appears to hide nothing, nothing but a green chaos at the core—which we brilliantly purposive apes can use and exploit as we please, with a free conscience.

A green chaos. Or a wood.

JOHN FOWLES, 'The Green Man', *The Tree*, 1979

The thing that first struck me about Bulow Hammock is the hardest to describe: the smell. Hammocks are woodlands (the name refers to hardwood groves that punctuate the more open marshes and pine woods of Florida, and may derive from Indian words for 'shady place,' 'garden place,' or 'floating plants'), but Bulow Hammock didn't smell like any woodlands I knew. I was used to the brisk, humus-and-chlorophyll tang of New England woods with their associations of uplifting weekend hikes. The hammock was different.

I must have been about nine years old when I encountered the hammock, so I didn't articulate any of this. Yet I clearly remember my sensations on stepping out of my parents' car into the shade of the magnolias and cabbage palmettoes. I was fascinated but daunted. The Connecticut woods I'd played in had been inviting, welcoming. The hammock was . . . seductive. It smelled sweet, a perfumey sweetness that reminded me of the hotel lobbies and cocktail lounges I'd occasionally been in with my parents.

Smells are hard to describe because we can't really remember them as we do sights and sounds, we can only recognize them. Smells lie deeper than our remembering, thinking neocortex, in the olfactory lobe we inherited from the early vertebrates. Yet smells are related to thought in profound ways because our nocturnal ancestors, the early mammals, lived by smell. The human ability to relate present to past and future may stem from their scent-tracking of food, an activity which takes place in time as well as space, unlike a hawk's immediate striking on sight, and thus implies planning. The curious resonance the olfactory senses have in memory, as when Proust tasted an epoch in a teacake, suggests that we have a great deal to learn from them.

Complex smells are the hardest to describe. Bulow Hammock smelled stranger than liquor and perfume. It smelled intricately spicy, with a sweetness not so much of flowers as of aromatic bark and leaves. There also was an air of decay in the sweetness; not the rich, sleepy, somewhat bitter decay of New England woods, but more of a nervous, sour atmosphere. When I scraped my foot over fallen leaves on the ground, I didn't uncover the soft brown dirt I was used to, but white sand and a network of fine, blackish roots like the hair of a buried animal. The sand was part of the smell too, a dusty, siliceous undertone to the spice and decay.

There was something dangerous about the smell, something inhibiting to my nine-year-old mind. I didn't want to rush into the hammock as I'd have wanted to rush into an unfamiliar Connecticut woodland. It wasn't that the hammock seemed ugly or repellent—on the contrary. The seductiveness was part of the inhibition. Perhaps it was just that the hammock was *so* unfamiliar. It's easy to read things into childhood memories. But the smell was powerful.

Society is suspicious of wild places because it fears a turning away from human solidarity toward a spurious, sentimental freedom. It is interesting, in this regard, to recall how *little* of freedom there was in my first perception of Bulow Hammock, how little of the unfettered feeling I got in sand dunes, hill meadows, pine woods or other open places that promised release from streets and classrooms. I wonder if the hammock inhibited me because there was more of humanity about it than a dune, meadow, or pine forest has; not of humanity in the sense of society and civilization, which (however irrationally, given the history of civilization) we associate with safety, but of animal humanity, of the walking primate that has spent most of its evolution in warm places like Florida: spicy, moldy, sandy places. Perhaps it wasn't the strangeness of the hammock that made it seem dangerously seductive, but a certain familiarity. It is, after all, dangerous to be human.

> DAVID RAINES WALLACE, 'The Green Tunnel', *Bulow Hammock*, 1988

Note for essay: If men could only disintegrate like autumn leaves, fret away, dropping their substance like chlorophyll, would not our attitude toward death be different? Suppose we saw ourselves burning like maples in a golden autumn.

> LOREN EISELEY, from *The Lost Notebooks of Loren Eiseley*, ed. Kenneth Heuer, 1987

I think about the valley. And it occurs to me more and more that everything I have seen is wholly gratuitous. The giant water bug's predations, the frog's croak, the tree with the lights in it are not in any real sense necessary per se to the world or to its creator. Nor am I. The creation in the first place, being itself, is the only necessity, for which I would die, and I shall. The point about that being, as I know it here and see it, is that, as I think about it, it accumulates in my mind as an extravagance of minutiae. The sheer fringe and network of detail assumes primary importance. That there are so many details seems to be the most important and visible fact about the creation. If you can't see the forest for the trees, then look at the trees; when you've looked at enough trees, you've seen a forest, you've got it. If the world is gratuitous, then the fringe of a goldfish's fin is a million times more so. The first question—the one crucial one—of the creation of the universe and the existence of something as a sign and an affront to nothing, is a blank one. I can't think about it. So it is to the fringe of that question that I affix my attention, the fringe of the fish's fin, the intricacy of the world's spotted and speckled detail.

The old Kabbalistic phrase is 'the Mystery of the Splintering

of the Vessels.' The words refer to the shrinking or imprisonment of essences within the various husk-covered forms of emanation or time. The Vessels splintered and solar systems spun;
ciliated rotifers whirled in still water, and newts with gills laid
tracks in the silt-bottomed creek. Not only did the Vessels splin-
ter: they splintered exceeding fine. Intricacy, then, is the subject,
the intricacy of the created world.

You are God. You want to make a forest, something to hold
the soil, lock up solar energy, and give off oxygen. Wouldn't it
be simpler just to rough in a slab of chemicals, a green acre of
goo?

You are a man, a retired railroad worker who makes replicas
as a hobby. You decide to make a replica of one tree, the longleaf pine your great-grandfather planted—just a replica—it doesn't
have to work. How are you going to do it? How long do you
think you might live, how good is your glue? For one thing, you
are going to have to dig a hole and stick your replica trunk in
the ground halfway to China if you want the thing to stand up.
Because you will have to work fairly big; if your replica is too
small, you'll be unable to handle the slender, three-sided needles,
affix them in clusters of three in fascicles, and attach those laden
fascicles to flexible twigs. The twigs themselves must be covered
by 'many silvery-white, fringed, long-spreading scales.' Are your
pine cones' scales 'thin, flat, rounded at the apex, the exposed
portions (closed cone) reddish brown, often wrinkled, armed on
the back with a small, reflexed prickle, which curves toward the
base of the scale'? When you loose the lashed copper wire truss-
ing the replica limbs to the trunk, the whole tree collapses like
an umbrella.

You are a starling. I've seen you fly through a longleaf pine
without missing a beat.

You are a sculptor. You climb a great ladder; you pour grease
all over a growing longleaf pine. Next, you build a hollow cylin-
der like a cofferdam around the entire pine, and grease its inside
walls. You climb your ladder and spend the next week pouring
wet plaster into the cofferdam, over and inside the pine. You
wait; the plaster hardens. Now open the walls of the dam, split
the plaster, saw down the tree, remove it, discard, and your intricate sculpture is ready: this is the shape of part of the air.

You are a chloroplast moving in water heaved one hundred
feet above ground. Hydrogen, carbon, oxygen, nitrogen in a ring

around magnesium. . . . You are evolution; you have only begun to make trees. You are God—are you tired? finished?

Intricacy means that there is a fluted fringe to the something that exists over against nothing, a fringe that rises and spreads, burgeoning in detail. Mentally reverse positive and negative space, as in the plaster cast of the pine, and imagine emptiness as a sort of person, a boundless person consisting of an elastic, unformed clay. (For the moment forget that the air in our atmosphere is 'something,' and count it as 'nothing,' the sculptor's negative space.) The clay man completely surrounds the holes in him, which are galaxies and solar systems. The holes in him part, expand, shrink, veer, circle, spin. He gives like water, he spreads and fills unseeing. Here is a ragged hole, our earth, a hole that makes torn and frayed edges in his side, mountains and pines. And here is the shape of one swift, raveling edge, a feather-hole on a flying goose's hollow wing extended over the planet. Five hundred barbs of emptiness prick into clay from either side of a central, flexible shaft. On each barb are two fringes of five hundred barbules apiece, making a million barbules on each feather, fluted and hooked in a matrix of clasped hollowness. Through the fabric of this form the clay man shuttles unerringly, and through the other feather-holes, and the goose, the pine forest, the planet, and so on.

In other words, even on the perfectly ordinary and clearly visible level, creation carries on with an intricacy unfathomable and apparently uncalled for. The lone ping into being of the first hydrogen atom *ex nihilo* was so unthinkably, violently radical, that surely it ought to have been enough, more than enough. But look what happens. You open the door and all heaven and hell break loose.

Evolution, of course, is the vehicle of intricacy. The stability of simple forms is the sturdy base from which more complex stable forms might arise, forming in turn more complex forms, and so on. The stratified nature of this stability, like a house built on rock on rock on rock, performs, in Jacob Bronowski's terms, as the 'ratchet' that prevents the whole shebang from 'slipping back.' Bring a feather into the house, and a piano; put a sculpture on the roof, sure, and fly banners from the lintels—the house will hold.

There are, for instance, two hundred twenty-eight separate and distinct muscles in the head of an ordinary caterpillar. Again, of an ostracod, a common fresh-water crustacean of the sort I crunch on by the thousands every time I set foot in Tinker Creek, I read, 'There is one eye situated at the fore-end of the animal. The food canal lies just below the hinge, and around the mouth are the feathery feeding appendages which collect the food. . . . Behind them is a foot which is clawed and this is partly used for removing unwanted particles from the feeding appendages.' Or again, there are, as I have said, six million leaves on a big elm. All right . . . but they are toothed, and the teeth themselves are toothed. How many notches and barbs is that to a world? In and out go the intricate leaf edges, and 'don't nobody know why.' All the theories botanists have devised to explain the functions of various leaf-shapes tumble under an avalanche of inconsistencies. They simply don't know, can't imagine.

ANNIE DILLARD, 'Intricacy', *Pilgrim at Tinker Creek*, 1975

If nature were about to end, we might muster endless energy to stave it off; if nature has already ended, what are we fighting for? Before any redwoods had been cloned or genetically improved, one could understand clearly what the fight against such tinkering was about. It was about the idea that a redwood was somehow sacred, that its fundamental identity should remain beyond our control. But once that barrier has been broken, what is the fight about then? It's not like opposing nuclear reactors or toxic waste dumps, each one of which poses new risks to new areas. This damage is to an idea, the idea of nature, and all the ideas that descend from it. It is not cumulative. Wendell Berry once argued that without a 'fascination' with the wonder of the natural world 'the energy needed for its preservation will never be developed'—that 'there must be a mystique of the rain if we are ever to restore the purity of the rainfall'. This makes sense when the problem is transitory—sulphur from a smoke-stack drifting over the Adirondacks. But how can there be a mystique of the rain now that every drop—even those drops that fall as snow on the Arctic, even those drops that fall deep in the remaining primeval forest—bears the permanent stamp of man? Having lost its separateness, it loses its special power. Instead of being a category like God—something beyond our control—it is a category like the defence budget or the dole, a problem we must work out. This in itself changes its meaning completely, and changes our reaction to it . . .

In the end, I understand perfectly well that defiance may mean prosperity, and a sort of security—that more dams will help the people of Phoenix, and that genetic engineering will help the sick, and that there is so much progress that can still be made against human misery. And I have no great desire to limit my way of life. If I thought we could put off the decision, foist it on our grandchildren, I'd be willing. As it is, I have no plans to live in a cave, or even an unheated cabin. If it took ten thousand years to get where we are, it will take a few generations to climb back down. But this could be the epoch when people decide at least to go no further down the path we've been following—when we make not only the necessary technological adjustments to preserve the world from overheating, but also the necessary mental adjustments to ensure that we'll never again put our good ahead of everything else's. This is the path I choose, for it offers at least a shred of hope for a living, meaningful world.

The reasons for my choice are as numerous as the trees on the hill outside my window, but they crystallized in my mind when I read a passage from one of the brave optimists of our managed future. 'The existential philosophers—particularly Sartre—used to lament that man lacked an essential purpose,' says Walter Truett Anderson. 'We find now that the human predicament is not quite so devoid of inherent purpose after all. To be caretakers of a planet, custodians of all its life forms and shapers of its (and our own) future is certainly purpose enough.' This intended rallying cry depresses me more deeply than I can say. That is our destiny? To be 'caretakers' of a managed world, 'custodians' of all life? For that job security we will trade the mystery of the natural world, the pungent mystery of our own lives and a world bursting with exuberant creation? Much better Sartre's neutral purposelessness. But, much better than that, another vision, of man actually living up to his potential.

As birds have flight, we have the special gift of reason. Part of that reason drives the intelligence that allows us, say, to figure out and master DNA, or build big power plants. But our reason could also keep us from following blindly the biological imperatives towards endless growth in numbers and territory. Our reason allows us to conceive of our species as a species, and to recognize the danger that our growth poses to it, and to feel something for the other species we threaten. Should we so choose, we could exercise our reason to do what no other animal can do:

we could limit ourselves voluntarily, *choose* to remain God's creatures instead of making ourselves gods. What a towering achievement that would be, so much more impressive than the largest dam (beavers can build dams), because so much harder. Such restraint—not genetic engineering or planetary management—is the real challenge, the hard thing. Of course we can splice genes. But can we *not* splice genes?

BILL McKIBBEN, The End of Nature, 1990

One swallow does not make a summer, but one skein of geese, cleaving the murk of a March thaw, is the spring.

A cardinal, whistling spring to a thaw but later finding himself mistaken, can retrieve his error by resuming his winter silence. A chipmunk, emerging for a sunbath but finding a blizzard, has only to go back to bed. But a migrating goose, staking two hundred miles of black night on the chance of finding a hole in the lake, has no easy chance for retreat. His arrival carries the conviction of a prophet who has burned his bridges.

A March morning is only as drab as he who walks in it without a glance skyward, ear cocked for geese. I once knew an educated lady, banded by Phi Beta Kappa, who told me that she had never heard or seen the geese that twice a year proclaim the revolving seasons to her well-insulated roof. Is education possibly a process of trading awareness for things of lesser worth? The goose who trades his is soon a pile of feathers.

The geese that proclaim the seasons to our farm are aware of many things, including the Wisconsin statutes. The south-bound November flocks pass over us high and haughty, with scarcely a honk of recognition for their favorite sandbars and sloughs. 'As a crow flies' is crooked compared with their undeviating aim at the nearest big lake twenty miles to the south, where they loaf by day on broad waters and filch corn by night from the freshly cut stubbles. November geese are aware that every marsh and pond bristles from dawn till dark with hopeful guns.

March geese are a different story. Although they have been shot at most of the winter, as attested by their buckshot-battered pinions, they know that the spring truce is now in effect. They wind the oxbows of the river, cutting low over the now gunless points and islands, and gabbling to each sandbar as to a long-lost friend. They weave low over the marshes and meadows, greeting each newly melted puddle and pool. Finally, after a few *pro-forma* circlings of our marsh, they set wing and glide silently to the pond, black landing-gear lowered and rumps white against the far hill. Once touching water, our newly arrived guests set up a honking and splashing that shakes the last thought of winter out of the brittle cattails. Our geese are home again! . . .

It is an irony of history that the great powers should have discovered the unity of nations at Cairo in 1943. The geese of the world have had that notion for a longer time, and each March they stake their lives on its essential truth.

In the beginning there was only the unity of the Ice Sheet. Then followed the unity of the March thaw, and the northward hegira of the international geese. Every March since the Pleistocene, the geese have honked unity from China Sea to Siberian Steppe, from Euphrates to Volga, from Nile to Murmansk, from Lincolnshire to Spitsbergen. Every March since the Pleistocene, the geese have honked unity from Currituck to Labrador, Matamuskeet to Ungava, Horseshoe Lake to Hudson's Bay, Avery Island to Baffin Land, Panhandle to Mackenzie, Sacramento to Yukon.

By this international commerce of geese, the waste corn of Illinois is carried through the clouds to the Arctic tundras, there to combine with the waste sunlight of a nightless June to grow goslings for all the lands between. And in this annual barter of food for light, and winter warmth for summer solitude, the whole continent receives as net profit a wild poem dropped from the murky skies upon the muds of March.

> ALDO LEOPOLD, 'The Geese Return', *A Sand County Almanac*, 1949

It is one of our problems that as we become crowded together, the sounds we make to each other, in our increasingly complex communication systems, become more random-sounding, accidental or incidental, and we have trouble selecting meaningful signals out of the noise. One reason is, of course, that we do not seem able to restrict our communication to information-bearing, relevant signals. Given any new technology for transmitting information, we seem bound to use it for great quantities of small talk. We are only saved by music from being overwhelmed by nonsense.

It is a marginal comfort to know that the relatively new science of bioacoustics must deal with similar problems in the sounds made by other animals to each other. No matter what sound-making device is placed at their disposal, creatures in general do a great deal of gabbling, and it requires long patience and observation to edit out the parts lacking syntax and sense. Light social conversation, designed to keep the party going, prevails. Nature abhors a long silence.

Somewhere, underlying all the other signals, is a continual music. Termites make percussive sounds to each other by beating their heads against the floor in the dark, resonating corridors of their nests. The sound has been described as resembling, to the human ear, sand falling on paper, but spectrographic analysis of sound records has recently revealed a high degree of organization in the drumming; the beats occur in regular, rhythmic phrases, differing in duration, like notes for a tympani section.

From time to time, certain termites make a convulsive movement of their mandibles to produce a loud, high-pitched clicking sound, audible ten meters off. So much effort goes into this one note that it must have urgent meaning, at least to the sender. He cannot make it without such a wrench that he is flung one or two centimeters into the air by the recoil.

There is obvious hazard in trying to assign a particular meaning to this special kind of sound, and problems like this exist throughout the field of bioacoustics. One can imagine a woolly-minded Visitor from Outer Space, interested in human beings, discerning on his spectrograph the click of that golf ball on the surface of the moon, and trying to account for it as a call of warning (unlikely), a signal of mating (out of the question), or an announcement of territory (could be).

Bats are obliged to make sounds almost ceaselessly, to sense, by sonar, all the objects in their surroundings. They can spot with accuracy, on the wing, small insects, and they will home onto things they like with infallibility and speed. With such a system for the equivalent of glancing around, they must live in a world of ultrasonic bat-sound, most of it with an industrial, machinery sound. Still, they communicate with each other as well, by clicks and high-pitched greetings. Moreover, they have been heard to produce, while hanging at rest upside down in the depths of woods, strange, solitary, and lovely bell-like notes.

Almost anything that an animal can employ to make a sound is put to use. Drumming, created by beating the feet, is used by prairie hens, rabbits, and mice; the head is banged by wood-peckers and certain other birds; the males of deathwatch beetles make a rapid ticking sound by percussion of a protuberance on the abdomen against the ground; a faint but audible ticking is made by the tiny beetle *Lepinotus inquilinus*, which is less than two millimeters in length. Fish make sounds by clicking their teeth, blowing air, and drumming with special muscles against tuned inflated air bladders. Solid structures are set to vibrating by toothed bows in crustaceans and insects. The proboscis of the death's-head hawk moth is used as a kind of reed instrument, blown through to make high-pitched, reedy notes.

Gorillas beat their chests for certain kinds of discourse. Animals with loose skeletons rattle them, or, like rattlesnakes, get sounds from externally placed structures. Turtles, alligators, crocodiles, and even snakes make various more or less vocal sounds. Leeches have been heard to tap rhythmically on leaves, engaging the attention of other leeches, which tap back, in synchrony. Even earth-worms make sounds, faint staccato notes in regular clusters. Toads sing to each other, and their friends sing back in antiphony.

Birdsong has been so much analyzed for its content of business communication that there seems little time left for music, but it is there. Behind the glossaries of warning calls, alarms, mating messages, pronouncements of territory, calls for recruit-ment, and demands for dispersal, there is redundant, elegant sound that is unaccountable as part of the working day. The thrush in my backyard sings down his nose in meditative, liquid runs of melody, over and over again, and I have the strongest impression that he does this for his own pleasure. Some of the time he seems to be practicing, like a virtuoso in his apartment. He starts a run, reaches a midpoint in the second bar where there should be a set of complex harmonics, stops, and goes back to begin over, dissatisfied. Sometimes he changes his notation so conspicuously that he seems to be improvising sets of variations. It is a meditative, questioning kind of music, and I cannot believe that he is simply saying, 'thrush here.'

The robin sings flexible songs, containing a variety of motifs that he rearranges to his liking; the notes in each motif consti-tute the syntax, and the possibilities of variation produce a con-siderable repertoire. The meadow lark, with three hundred notes to work with, arranges these in phrases of three to six notes and elaborates fifty types of song. The nightingale has twenty-four basic songs, but gains wild variety by varying the internal arrangement of phrases and the length of pauses. The chaffinch listens to other chaffinches, and incorporates into his memory snatches of their songs.

The need to make music, and to listen to it, is universally expressed by human beings. I cannot imagine, even in our most primitive times, the emergence of talented painters to make cave paintings without there having been, near at hand, equally creative people making song. It is, like speech, a dominant aspect of human biology.

The individual parts played by other instrumentalists—crickets or earthworms, for instance—may not have the sound of music by themselves, but we hear them out of context. If we could listen to them all at once, fully orchestrated, in their immense ensemble, we might become aware of the counterpoint, the balance of tones and timbres and harmonics, the sonorities. The recorded songs of the humpback whale, filled with tensions and resolutions, ambiguities and allusions, incomplete, can be listened to as a *part* of music, like an isolated section of an orchestra. If we had better hearing, and could discern the descants of sea birds, the rhythmic tympani of schools of mollusks, or even the distant harmonics of midges hanging over meadows in the sun, the combined sound might lift us off our feet.

There are, of course, other ways to account for the songs of whales. They might be simple, down-to-earth statements about navigation, or sources of krill, or limits of territory. But the proof is not in, and until it is shown that these long, convoluted, insistent melodies, repeated by different singers with ornamentations of their own, are the means of sending through several hundred miles of undersea such ordinary information as 'whale here,' I shall believe otherwise. Now and again, in the intervals between songs, the whales have been seen to breach, leaping clear out of the sea and landing on their backs, awash in the turbulence of their beating flippers. Perhaps they are pleased by the way the piece went, or perhaps it is celebration at hearing one's own song returning after circumnavigation; whatever, it has the look of jubilation.

I suppose that my extraterrestrial Visitor might puzzle over my records in much the same way, on first listening. The 14th Quartet might, for him, be a communication announcing, 'Beethoven here,' answered, after passage through an undersea of time and submerged currents of human thought, by another long signal a century later, 'Bartok here.'

If, as I believe, the urge to make a kind of music is as much a characteristic of biology as our other fundamental functions, there ought to be an explanation for it. Having none at hand, I am free to make one up. The rhythmic sounds might be the recapitulation of something else—an earliest memory, a score for the transformation of inanimate, random matter in chaos into the improbable, ordered dance of living forms . . .

If there were to be sounds to represent this process, they would have the arrangement of the Brandenburg Concertos for my ear, but I am open to wonder whether the same events are recalled by the rhythms of insects, the long, pulsing runs of birdsong, the descants of whales, the modulated vibrations of a million locusts in migration, the tympani of gorilla breasts, termite heads, drumfish bladders. A 'grand canonical ensemble' is, oddly enough, the proper term for a quantitative model system in thermodynamics, borrowed from music by way of mathematics. Borrowed back again, provided with notation, it would do for what I have in mind.

LEWIS THOMAS, 'The Music of This Sphere', The Lives of a Cell, 1974

Our character lies for hundreds of millions of years, bound to three atoms of oxygen and one of calcium, in the form of limestone: it already has a very long cosmic history behind it, but we shall ignore it. For it time does not exist, or exists only in the form of sluggish variations in temperature, daily or seasonal, if, for the good fortune of this tale, its position is not too far from the earth's surface. Its existence, whose monotony cannot be thought of without horror, is a pitiless alternation of hots and colds, that is, of oscillations (always of equal frequency) a trifle more restricted and a trifle more ample: an imprisonment, for this potentially living personage, worthy of the Catholic Hell. To it, until this moment, the present tense is suited, which is that of description, rather than the past tense, which is that of narration—it is congealed in an eternal present, barely scratched by the moderate quivers of thermal agitation.

But, precisely for the good fortune of the narrator, whose story could otherwise have come to an end, the limestone rock ledge of which the atom forms a part lies on the surface. It lies within reach of man and his pickax (all honor to the pickax and its modern equivalents; they are still the most important intermediaries in the millennial dialogue between the elements and man): at any moment—which I, the narrator, decide out of pure caprice to be the year 1840—a blow of the pickax detached it and sent it on its way to the lime kiln, plunging it into the world of things that change. It was roasted until it separated from the calcium, which remained so to speak with its feet on the ground and went to meet a less brilliant destiny, which we shall not narrate. Still firmly clinging to two of its three former oxygen companions, it issued from the chimney and took the path of the air. Its story, which once was immobile, now turned tumultuous.

It was caught by the wind, flung down on the earth, lifted ten kilometers high. It was breathed in by a falcon, descending into its precipitous lungs, but did not penetrate its rich blood and was expelled. It dissolved three times in the water of the sea, once in the water of a cascading torrent, and again was expelled. It traveled with the wind for eight years: now high, now low, on the sea and among the clouds, over forests, deserts, and limitless expanses of ice; then it stumbled into capture and the organic adventure . . .

Now our atom is inserted: it is part of a structure, in an architectural sense; it has become related and tied to five companions so identical with it that only the fiction of the story permits me to distinguish them. It is a beautiful ring-shaped structure, an almost regular hexagon, which however is subjected to complicated exchanges and balances with the water in which it is dissolved; because by now it is dissolved in water, indeed in the sap of the vine, and this, to remain dissolved, is both the obligation and the privilege of all substances that are destined (I was about to say 'wish') to change. And if then anyone really wanted to find out why a ring, and why a hexagon, and why soluble in water, well, he need not worry: these are among the not many questions to which our doctrine can reply with a persuasive discourse, accessible to everyone, but out of place here.

It has entered to form part of a molecule of glucose, just to speak plainly: a fate that is neither fish, flesh, nor fowl, which is intermediary, which prepares it for its first contact with the animal world but does not authorize it to take on a higher responsibility: that of becoming part of a proteic edifice. Hence it travels, at the slow pace of vegetal juices, from the leaf through the pedicel and by the shoot to the trunk, and from here descends to the almost ripe bunch of grapes. What then follows is the province of the winemakers: we are only interested in pinpointing the fact that it escaped (to our advantage, since we would not know how to put it in words) the alcoholic fermentation, and reached the wine without changing its nature.

It is the destiny of wine to be drunk, and it is the destiny of glucose to be oxidized. But it was not oxidized immediately: its drinker kept it in his liver for more than a week, well curled up and tranquil, as a reserve aliment for a sudden effort; an effort that he was forced to make the following Sunday, pursuing a bolting horse. Farewell to the hexagonal structure: in the space of a few instants the skein was unwound and became glucose again, and this was dragged by the bloodstream all the way to a minute muscle fiber in the thigh, and here brutally split into two molecules of lactic acid, the grim harbinger of fatigue: only later, some minutes after, the panting of the lungs was able to supply the oxygen necessary to quietly oxidize the latter. So a new molecule of carbon dioxide returned to the atmosphere, and a parcel of the energy that the sun had handed to the vine-shoot passed from the state of chemical energy to that of mechanical energy, and thereafter settled down in the slothful condition of heat, warming up imperceptibly the air moved by the running and the blood of the runner. 'Such is life,' although rarely is it described in this manner: an inserting itself, a drawing off to its advantage, a parasitizing of the downward course of energy, from its noble solar form to the degraded one of low-temperature heat. In this downward course, which leads to equilibrium and thus death, life draws a bend and nests in it.

Our atom is again carbon dioxide, for which we apologize: this too is an obligatory passage; one can imagine and invent others, but on earth that's the way it is. Once again the wind, which this time travels far; sails over the Apennines and the Adriatic, Greece, the Aegean, and Cyprus: we are over Lebanon, and the dance is repeated. The atom we are concerned with is now trapped in a structure that promises to last for a long time: it is the venerable trunk of a cedar, one of the last; it is passed again through the stages we have already described, and the glucose of which it is a part belongs, like the bead of a rosary, to a long chain of cellulose. This is no longer the hallucinatory and geological fixity of rock, this is no longer millions of years, but we can easily speak of centuries because the cedar is a tree of great longevity. It is our whim to abandon it for a year or five hundred years: let us say that after twenty years (we are in 1868) a wood worm has taken an interest in it. It has dug its tunnel between the trunk and the bark, with the obstinate and blind voracity of its race; as it drills it grows, and its tunnel grows with it. There it has swallowed and provided a setting for the subject of this story; then it has formed a pupa, and in the spring it has come out in the shape of an ugly gray moth which is now drying in the sun, confused and dazzled by the splendor of the day. Our atom is in one of the insect's thousand eyes, contributing to the summary and crude vision with which it orients itself in space. The insect is fecundated, lays its eggs, and dies: the small cadaver lies in the undergrowth of the woods, it is emptied of its fluids, but the chitin carapace resists for a long time, almost indestructible. The snow and sun return above it without injuring it: it is buried by the dead leaves and the loam, it has become a slough, a 'thing,' but the death of atoms, unlike ours, is never irrevocable. Here are at work the omnipresent, untiring, and invisible gravediggers of the undergrowth, the microorganisms of the humus. The carapace, with its eyes by now blind, has slowly disintegrated, and the ex-drinker, ex-cedar, ex-wood worm has once again taken wing . . .

It is again among us, in a glass of milk. It is inserted in a very complex, long chain, yet such that almost all of its links are acceptable to the human body. It is swallowed; and since every living structure harbors a savage distrust toward every contribution of any material of living origin, the chain is meticulously broken apart and the fragments, one by one, are accepted or rejected. One, the one that concerns us, crosses the intestinal threshold and enters the bloodstream: it migrates, knocks at the door of a nerve cell, enters, and supplants the carbon which was part of it. This cell belongs to a brain, and it is my brain, the brain of the *me* who is writing; and the cell in question, and within it the atom in question, is in charge of my writing, in a gigantic minuscule game which nobody has yet described. It is that which at this instant, issuing out of a labyrinthine tangle of yeses and nos, makes my hand run along a certain path on the paper, mark it with these volutes that are signs: a double snap, up and down, between two levels of energy, guides this hand of mine to impress on the paper this dot, here, this one.

PRIMO LEVI, 'Carbon', *The Periodic Table*, 1985

Acknowledgements

The editor and publisher are grateful for permission to include the following copyright material:

Kenneth Allsop, from *The Sunday Times Magazine*, 23 March 1969, © Kenneth Allsop 1969. Used with permission.

Anon., Irish, 'May-time', trans. Kenneth Jackson, from *A Celtic Miscellany* (Penguin Books, 1971), translation © 1971 Kenneth Jackson. Used with permission.

J. A. Baker, from *The Peregrine* (Collins, 1967).

H. E. Bates, from *Through the Woods* (Victor Gollancz Ltd., 1936), © The Estate of H. E. Bates. Used with permission.

Jocelyn Brooke, from *The Military Orchid* (Bodley Head, 1948).

Giraldus Cambrensis, version by Geoffrey Grigson, reprinted in *Rainbows, Fleas and Flowers* (John Baker, 1971). Reprinted by permission of David Higham Associates Ltd.

Tai Ping Kuang Chi, translation © Jorge Luis Borges. Used with permission of Georges Borchardt Inc.

Garth Christian, from *Down the Long Wind* (George Newnes, 1963).

Colette, from *Prisons et Paradis* (1932) and *En Pays Connu* (1950), trans. Derek Coltman, in *An Earthly Paradise* (1966). Reprinted by permission of Secker & Warburg Ltd.

John Stewart Collis, from *Down to Earth* (1947).

Cyril Connolly, from *The Condemned Playground* (1945). Reprinted by permission of Deborah Rogers Ltd.

The Earl of Cranbrook, reprinted in *The Countryman Wildlife Book* (David & Charles, 1969). Used with permission.

Jacques Delamain, from *Why Birds Sing*, trans. Ruth & Anna Sarason (Victor Gollancz Ltd.). Used with permission.

Annie Dillard, from *Pilgrim at Tinker Creek* (Cape, 1975).

Loren Eiseley, from *The Lost Notebooks of Loren Eiseley*, ed. Kenneth Heuer (Little Brown, 1987). Used with permission.

Gretel Ehrlich, from *Antaeus on Nature*, ed. Daniel Halpern (Ecco Press, 1986), © Gretel Ehrlich 1986.

Patrick Leigh Fermor, from *Between the Woods and the Water* (1986). Reprinted by permission of John Murray (Publishers) Ltd.

Chris Ferris, from *The Darkness is Light Enough* (Michael Joseph, 1986). Used with permission.

John Fowles, from *The Tree*, © John Fowles 1979 (Little Brown & Co.). Used with permission.

James Fisher, from *The Shell Bird Book* (Ebury Press & Michael Joseph, 1966). Used with permission.

Jane Goodall, from *Through a Window: 30 Years with the Chimpanzees of Gombe* (Weidenfeld & Nicholson, 1990). Used with permission.

Geoffrey Grigson, from *The Englishman's Flora* (Phoenix House, 1958); from *A Herbal of All Sorts* (Phoenix House, 1959). Reprinted by permission of David Higham Associates Ltd.

Gruffyd ab Addaf ap Dafydd, trans. Kenneth Jackson, in *A Celtic Miscellany* (Penguin, 1971). Trans. © 1971 Kenneth Jackson. Used with permission.

J. B. S. Haldane, from *Possible Worlds* (Chatto & Windus, 1975). Used with permission.

Bernd Heinrich, from *Raven in Winter* (Barrie & Jenkins, 1990). Used with permission.

Edward Hoagland, from *Hearts Desire* (Collins Harvill, 1990). Used with permission.

Thomas Johnson, trans. C. E. Raven et al. in *Thomas Johnson: Botanical Journeys In Kent and Hampstead*, ed. J. S. L. Gilmour, © Hunt Botanical Library, Pittsburgh, 1972. Used with permission.

Maxwell Knight, first published in *The Countryman* magazine, reprinted in *The Countryman Wildlife Book* (David & Charles, 1969). Used with permission.

Aldo Leopold, from *A Sand County Almanac* (Oxford University Press, 1949). Used with permission.

Edward Lhwyd, from *Life and Letters of Edward Lhwyd*, 'Early Science in Oxford', Vol. XIV, 1945.

Carl Linnaeus, from *Journals*, quoted and trans. by Wilfrid Blunt, in *The Compleat Naturalist* (Collins, 1971).

Peter Matthiessen, from *The Tree Where Man Was Born* (Collins, 1972, an imprint of HarperCollins Publishers, Ltd.). Used with permission.

Gavin Maxwell, from *Ring of Bright Water*, © Gavin Maxwell 1960. Reprinted by permission of Longman Group UK Ltd.

Primo Levi, from *The Periodical Table* (Michael Joseph, 1985), © Primo Levi 1985. Used with permission.

Barry Lopez, from *Arctic Dreams* (Macmillan Press Ltd., 1986). Used with permission.

Margaret Mee, from *Flowers of the Amazon Forests* (Nonesuch Expeditions, 1988), © Margaret Mee 1988.

Norman Nicholson, from *Second Nature* (Cape, 1984). Reprinted by permission of David Higham Associates Ltd.

George Orwell, from *The Collected Essays of George Orwell*, Vol. IV (1968), © 1968 The Estate of George Orwell. Reprinted by permission of A. M. Heath Ltd., and Harcourt Brace Jovanovich on behalf of Sonia Brownell Orwell.

Oliver Rackham, from *The History of the Countryside* (J. M. Dent & Sons Ltd., 1986). Used with permission.

John Ray, from *Ray's Flora of Cambridgeshire*, trans. and ed. A. H. Ewen and C. T. Prince (Wheldon & Wesley Ltd., 1975). Used with permission.

Lewis Thomas, from *The Lives of a Cell* (Penguin, 1974), © Lewis Thomas 1974. Used with permission.

Niko Tinbergen, from *Curious Naturalists* (Penguin, 1974), © Niko Tinbergen 1974. Used with permission.

William Turner, trans. in Charles E. Raven, *English Naturalists From Neckham to Ray* (Cambridge University Press, 1947).

David Raines Wallace, from *The Green Tunnel* (Bulow Hammock, Sierra Club, 1988).

E. B. White, from *Essays of E. B. White* (Harper & Row, 1977). First appeared in *The New Yorker*, 1956, © The New Yorker Magazine Inc., 1956, 1977. Used with permission.

Edward O. Wilson, reprinted by permission of the President and Fellows of Harvard University, Cambridge, Mass., from *The Diversity of Life*, © The Belknap Press, Harvard University Press, 1992.

Andrew Young, from *A Prospect of Flowers* (Cape, 1945). Reprinted by permission of Alison Lowbury, Literary Executor.

While every effort has been made to secure permission, we may have failed in a few cases to trace the copyright holder. We apologize for any apparent negligence.

Index of Authors

General Index